INTERMEDIATE ALGEBRA, 5TH EDITION, BY WOOTON AND DROOYAN

SOLUTIONS MANUAL FOR STUDENTS

BERNARD FELDMAN
Los Angeles Pierce College

Wadsworth Publishing Company
Belmont, California
A division of Wadsworth, Inc.

Mathematics Editor: Richard Jones
Typist: Betty Ritter
Copy Editor: Charles Cox

ISBN 0-534-00739-2

Printed in the United States of America

1 2 3 4 5 6 7 8 9 10---83 82 81 80

PREFACE

This student supplement has been prepared to accompany the 5th edition of *Intermediate Algebra*. Detailed solutions, including graphs, are included for the even-numbered exercises for each section.

Many of the exercises can be solved in different ways. In order to keep this supplement to a convenient size, alternate solutions are in general not shown. It is anticipated that sometimes you will devise a way to solve a problem that is simpler than the way shown. However, if you have difficulty with a particular exercise, you should refer to the solution given to see one way in which it can be solved. Of course, it is best for you to do this only after you have made a diligent attempt to work the problem yourself.

Bernard Feldman

CONTENTS

1

THE SET OF REAL NUMBERS

EXERCISE 1.1

2. {0,1,2,3}
4. {-1,0,1}
6. {3,5,7,9,11}
8. {11,22,23, . . .}
10. {whole numbers (or natural numbers or integers) between 6 and 10}
12. {natural number (or whole or integral) mutliples of 4}
14. {even whole numbers (or natural numbers or integers) between 11 and 17}
16. {8}; 8 is the only member of A that is a natural number.
18. $\left\{-5,-3.44, \ldots, -\frac{2}{3}, 0, \frac{1}{5}, \frac{7}{3}, 6.1, 8\right\}$; $\sqrt{15}$ is the only member in A that is not a rational number.
20. $\{x \mid x \in A\}$; all the members in A are real numbers.
22. Finite
24. Infinite; consider 0, 10, 100, 1000, . . .
26. Infinite; there are infinitely many real numbers between *any* two real numbers.
28. Variable, since there is more than one whole number.

30. Constant; there is only *one* natural number less than 2.

32. Variable, since there is more than one rational number between 0 and 1.

34. Since $\{0\}$ and {integers} are both sets of numbers and 0 is an integer, $\{0\} \subset$ {integers}.

36. Since $\{1,2\}$ and $\{1,2\}$ are both sets and 1 and 2 are members of each, $\{1,2\} \subset \{1,2\}$.

38. Since 0 is an *element* of $\{0\}$, $0 \in \{0\}$.

40. 3 and 4 are both natural numbers; hence, $\{3,4\} \subset$ {natural numbers}.

42. The null set is a subset of every set; hence, $\emptyset \subset$ {whole numbers}.

44. 0 is not a natural number; hence, $0 \notin$ {natural numbers}.

46. -3 is not a whole number; hence, $-3 \notin$ {whole numbers}.

48. {integers} $\not\subset$ {natural numbers}

50. $\{5\} \notin \{3,4,5\}$

52. {integers} $\subset$ {rational numbers}

54. {irrational numbers} $\subset$ {real numbers}

56. No. Since $A \neq B$, B will have members which A does not have.

58. No. Consider: $A = \{1,2\}$, $B = \{1\}$, $C = \{1,2\}$; $A \neq B$ and $B \neq C$, yet $A = C$.

60. Yes. Since $R \subset S$, every element of R is an element of S.

62. No. Consider $S = \{a\}$ and $T = \{a,b\}$.

64. $\emptyset$, $\{0\}$, $\{1\}$, $\{2\}$, $\{3\}$, $\{0,1\}$, $\{0,2\}$, $\{0,3\}$, $\{1,2\}$, $\{1,3\}$, $\{2,3\}$, $\{0,1,2\}$, $\{0,1,3\}$, $\{0,2,3\}$, $\{1,2,3\}$, $\{0,1,2,3\}$

66. 10. Consider the five-member set $\{a,b,c,d,e\}$. It is necessary to count how many two-member subsets it has. They are: $\{a,b\}$, $\{a,c\}$, $\{a,d\}$, $\{a,e\}$, $\{b,c\}$, $\{b,d\}$, $\{b,e\}$, $\{c,d\}$, $\{c,e\}$, $\{d,e\}$.

EXERCISE 1.2

2. $t + 3$
4. 3
6. a
8. 4
10. $2 + x$
12. $-3 < 3$
14. $-20 > -29$
16. $t - 4 \leqslant 0$ or $t - 4 \ngtr 0$
18. $x + 4 < 0$
20. $x \not\geqslant 4$
22. $0 \leqslant 3t \leqslant 4$
24. $<$
26. $>$
28. $<$
30. $=$
32. $>$; $>$
34. $<$; $<$
36. 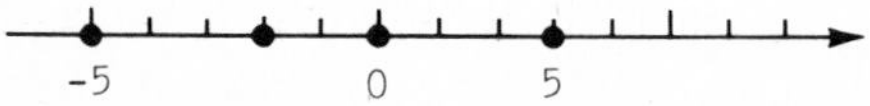
38. 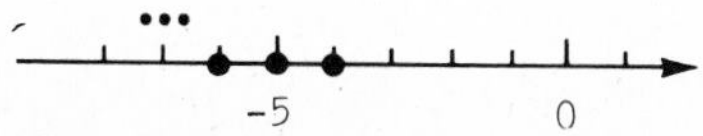 The 3 dots above the number line mean that the graph continues indefinitely to the left.
40. 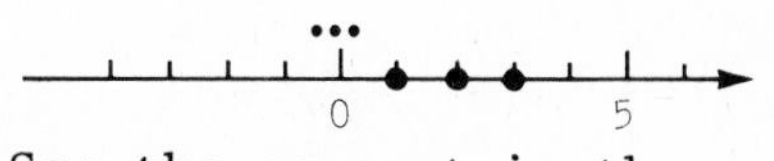

See the comment in the solution to Problem 38.
42. 
44. $-1 > -2$
46. $-3 < 0$
48. $x \leqslant z$
50. $x \not\geqslant 0$, $x < 0$
52. $x \ngtr 0$, $x \leqslant 0$

EXERCISE 1.3

2. 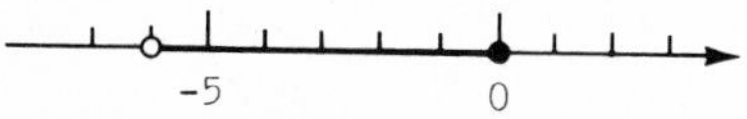
4. 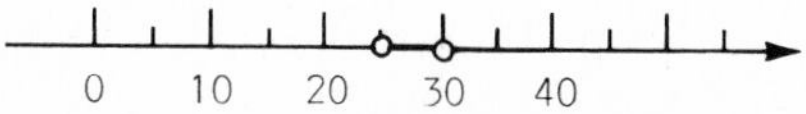
6. 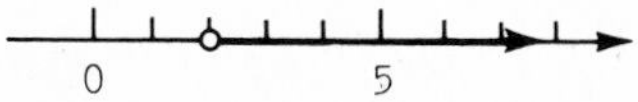
8. 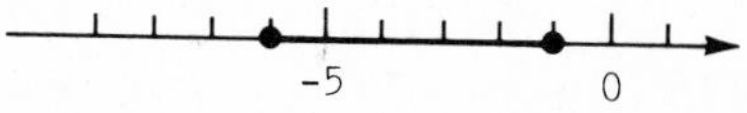
10. 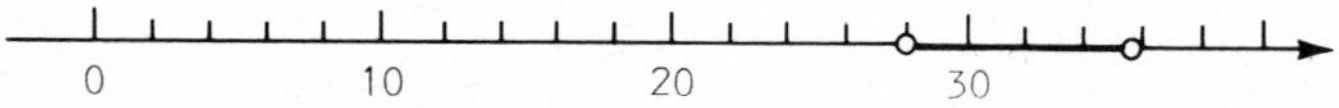
12. 
14. 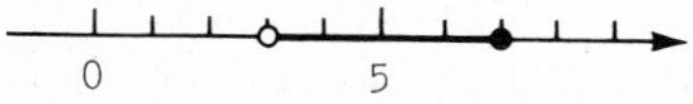 (3,7]

16. 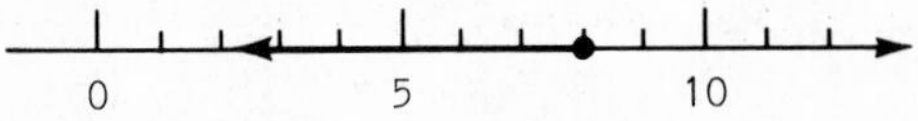

$(-\infty, 8]$

18. 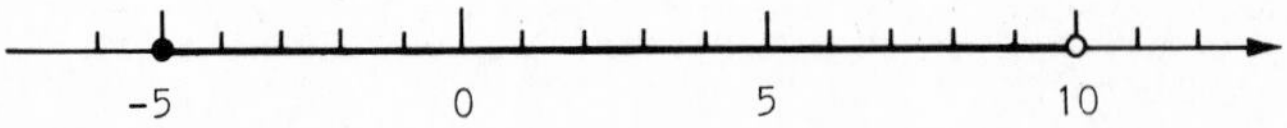

$[-5, 10)$

20. 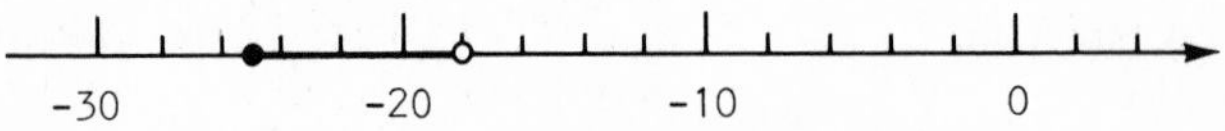

$[-25, -18)$

22. 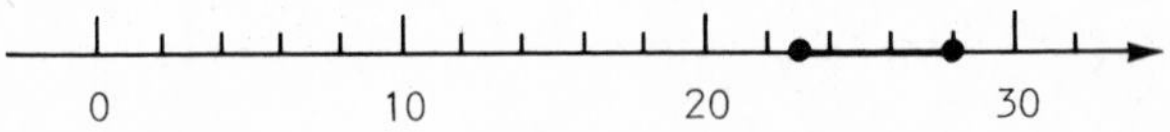

$[23, 28]$

24. 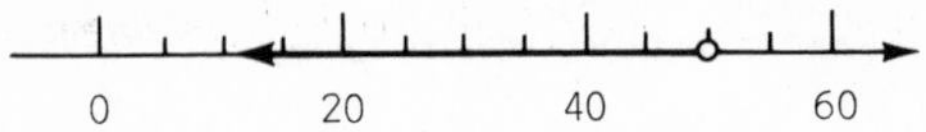

$(-\infty, 50)$

26. 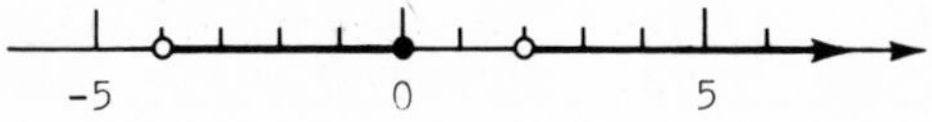

$\{x \mid -4 < x \leqslant 0\}$; $\{x \mid x > 2\}$

28. 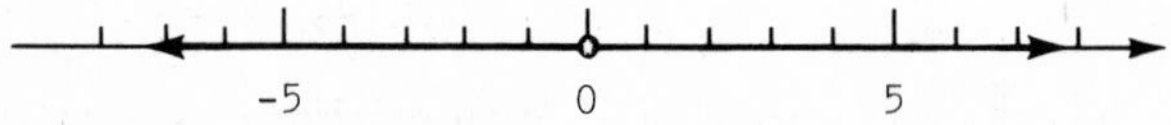

$\{x \mid x < 0\}$; $\{x \mid x > 0\}$

30. 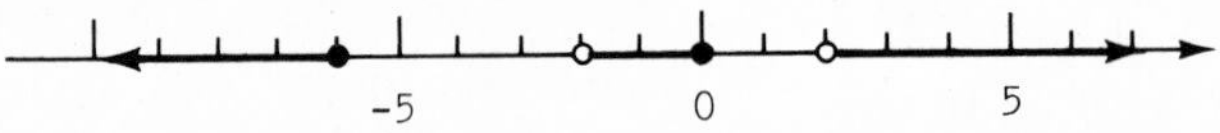

$\{x \mid x \leqslant -4\}$; $\{x \mid -2 < x \leqslant 0\}$; $\{x \mid x > 2\}$

EXERCISE 1.4

2. $(2m)n = 2(mn)$

4. $(2 + z) + 3 = 2 + (z + 3)$

6. (-7)

8. 0

10. r

12. 1

14. 3

16. 0

18. t; 3

20. Negative number; because $-(-x) = x$.

22. 18

24. 12

26. $|-4| = 4$; therefore, $-|-4| = -4$

28. $-\frac{3}{4}$

30. $2z$, if $z \geqslant 0$; $-2z$, if $z < 0$.

32. $-|-x|$ is nonpositive; therefore, $-|-x| = -x$ if $x \geqslant 0$; x, if $x < 0$.

34. $x - 5$, if $x - 5 \geqslant 0$ or $x \geqslant 5$; $-(x - 5)$, if $x - 5 < 0$ or $x < 5$.

36. $x + 3$, if $x + 3 \geqslant 0$ or $x \geqslant -3$; $-(x + 3)$, if $x + 3 < 0$ or if $x < -3$.

38. Not closed with respect to addition, since $1 + 1 = 2$ and $2 \notin \{-1,0,1\}$. Closed with respect to multiplication, since $(-1) \cdot (-1)$, $-1 \cdot 0$, $0 \cdot 0$, $0 \cdot 1$, and $1 \cdot 1$ are all elements of $\{-1,0,1\}$.

40. All of the sets are closed with respect to addition and with respect to multiplication.

42. No. If $x = 0$, they are not equal. If $x \neq 0$, the left member denotes a positive number while the right member denotes a negative number.

44. Yes. If $x \neq 0$ and $y \neq 0$, $|y|$ will always be less than $|x| + |y|$. Consider, for example, $x = -2$ and $y = 1$: $|1| < |-2| + |1|$ is true.

46. No. Consider $x = -2$ and $y = 1$. Then $|-2 + 1| \not> |-2|$.

48. $|y + 1| = |y - 1|$ if $y + 1 = y - 1$, which is impossible, or if $y + 1 = -(y - 1)$, which is possible only if $y = 0$.

EXERCISE 1.5

2. $8 + (-1)$

4. $-8 + (-3)$

6. $3 + [-(-5)]$ or $3 + 5$

8. $1 + [-(-1)] = 1 + 1 = 2$

10. 30

12. $-(|-7|-2) = -5$

14. $-(|-11|-6) = -5$

16. $-(|-7| + |-5|) = -12$

18. 4

20. $7 + (-18) = -(|-18|-7) = -11$

22. $-6 + (-12) = -(|-6| + |-12|) = -18$

24. $7 + [-(-2)] = 9$

26. $-11 + [-(-7)] = -11 + 7$
$= -(|-11|-7) = -4$

28. $(11 - 5) + 7 = 6 + 7$
$= 13$

30. $(4 - 3) - 11 = 1 - 11$
$= -10$

32. $(8 + [-(-2)]) + (-3)$
$= (10) + (-3) = 7$

34. $8 - (8) = 0$

36. $[3 + (-7)] + (-1)$
$= [-4] + (-1) = -5$

38. $4 - ([6 + 2] + [-11])$
$= 4 - ([8] + [-11])$
$= 4 - (-3) = 4 + 3 = 7$

40. $[5 + (-9)] + (2)$
$= [-4] + (2)$
$= -2$

42. $(13) + (13 + [(-17) + (-2)]) = 13 + (13 + [-19])$
$= 13 + (-6)$
$= 7$

44. $(-15 + [3 + (-1)]) - [(9 - 5) + 12]$
$= (-15 + [2]) - [(4) + 12]$
$= (-13) - (16)$
$= (-13) + (-16) = -29$

46. $[(-27 + 3) + (-6)] - [19 + (-12)] = [-24 + (-6)] - [7]$
$= -30 + (-7) = -37$

48. Since $(4 - 3) - 2 = 1 - 2 = -1$ and $4 - (3 - 2) = 4 - 1 = 3$, it follows that $(4 - 3) - 2 \neq 4 - (3 - 2)$. Hence, subtraction is not associative.

EXERCISE 1.6

2. -24 4. 30 6. $[(-3)(2)](6) = [-6](6) = -36$

8. $[(4)(-3)](-6) = [-12](-6) = 72$

10. $[(-8)(-3)](-1) = [24](-1) = -24$

12. $[(-6)(5)](0) = [-30](0) = 0$

14. $[(-1)(-2)][(-3)(-4)] = [2][12] = 24$

16. $[(7)(-5)][(-3)(-1)] = [-35][3] = -105$

18. $[(-2)(-5)][(2)(-1)(-1)] = [10][2] = 20$

20. $[(-4)(1)][(-1)(-2)(-3)] = [-4][-6] = 24$

22. -4 24. -3 26. 9 28. 0

30. Undefined 32. -4 34. $46 = (-23)(-2)$

36. $-14 = 2(-7)$ 38. $-28 = (-7)(4)$

40. $-12 = 1(-12)$ 42. $27\left(\frac{1}{7}\right)$

44. $9\left(\frac{1}{17}\right)$ 46. $3\left(\frac{1}{13}\right)$ 48. $3\left(\frac{1}{1000}\right)$ 50. $\frac{8}{3}$

52. $\frac{3}{100}$ 54. $\frac{6}{7}$ 56. $\frac{7}{10{,}000}$

58. Let $x = 1$ and $y = 2$. Then $-3(xy) = -3[(1)(2)] = -6$ and $(-3x)(-3y) = [(-3)(1)][(-3)(2)] = 18$. Therefore, $-3(xy)$ and $(-3x)(-3y)$ are not equal for all x and y.

60. True if $x \neq 0$ or if $y = 0$. A fraction equals zero only if its numerator equals zero. Also, a fraction is undefined if its denominator equals zero.

62. True if $x > 0$ and $y < 0$ or if $x < 0$ and $y > 0$. A fraction is negative if its numerator and denominator have opposite signs.

64. Since $(8 \div 4) \div 2 = 2 \div 2 = 1$ and $8 \div (4 \div 2) = 8 \div 2 = 4$,
$$(8 \div 4) \div 2 \neq 8 \div (4 \div 2).$$
Hence, division is not associative.

66. a. Since $1/2$ is division involving natural numbers and the quotient $1/2$ is not a natural number, the natural numbers are not closed under division.

b. Not closed. See part a above. 1 and 2 are whole numbers.

c. Not closed. See part a above. 1 and 2 are integers.

d. Closed. As long as divisors are not zero, every quotient of rational numbers is a rational number.

e. Closed. As long as divisors are not zero, every quotient of real numbers is a real number.

EXERCISE 1.7

2. $7 - 12 = -5$

4. $18 + (-10) = 8$

6. $5(-6) + 7 = -30 + 7$
$= -23$

8. $4 - 7(10) = 4 - 70$
$= -66$

10. $\frac{12}{6} - 5 = 2 - 5$
$= -3$

12. $\frac{5(-2)}{2} - \frac{18}{-3} = \frac{-10}{2} - (-6)$
$= -5 + 6 = 1$

14. $6[5 - 3(-3)] + 3$
$= 6[5 + 9] + 3$
$= 6[14] + 3$
$= 84 + 3 = 87$

16. $(2)[5 + 7(-1)]$
$= (2)[5 - 7]$
$= (2)[-2]$
$= -4$

18. $27 \div (3[9 - 3(2)])$
$= 27 \div (3[9 - 6])$
$= 27 \div (3[3])$
$= 27 \div (9) = 3$

20. $-3[-2 + 5] \div [-9]$
$= -3[3] \div [-9]$
$= -9 \div [-9]$
$= 1$

22. $[-2 + 3(-3)] \cdot [-15 \div 3]$
$= [-2 + (-9)] \cdot [-5]$
$= [-11] \cdot [-5] = 55$

24. $\frac{12 + 2}{2} = \frac{14}{2}$
$= 7$

26. $\frac{4 - 2(12)}{1 - 2} = \frac{4 - 24}{-1}$
$= 20$

28. $2.5[1 + (0.05)(20)]$
$= 2.5[1 + 1]$
$= 2.5[2] = 5.0$

30. $\frac{20}{2}(6 + 12) = 10(18)$
$= 180$

32. $\left[\frac{10}{-5}\right]\left[\frac{-9}{3}\right] = [-2][-3]$
$= 6$

34. $\left[7 + 3\left(\frac{-12}{6}\right) - 5\right] + 3$
$= [7 + 3(-2) - 5] + 3$
$= [(7 - 6) - 5] + 3$
$= [1 - 5] + 3$
$= [-4] + 3$
$= -1$

36. $\dfrac{12 + 3\left(\frac{-8}{2}\right) - 1}{-8 + 6\left(\frac{-18}{-3}\right) + 1}$
$= \dfrac{12 + 3(-4) - 1}{-8 + 6(6) + 1}$
$= \dfrac{[12 + (-12)] - 1}{-8 + (36 + 1)}$
$= \dfrac{0 - 1}{-8 + 37} = \dfrac{-1}{29}$

38. $$\frac{6 - 2\left(\frac{10}{5}\right) + 8}{3 - 6 + 8} = \frac{6 - 2(2) + 8}{-3 + 8}$$

$$= \frac{(6 - 4) + 8}{5}$$

$$= \frac{2 + 8}{5}$$

$$= \frac{10}{5}$$

$$= 2$$

40. Let $y = 0$. Then

$$7 - 2 \cdot y = 7 - 2(0) = 7$$

and

$$(7 - 2)y = (7 - 2)(0) = 0.$$

Therefore, $7 - 2 \cdot y \neq (7 - 2)y$ for all y.

2

POLYNOMIALS

EXERCISE 2.1

2. Monomial; degree 5

4. Trinomial; degree 2 in z

6. Trinomial; degree 2

8. Two terms

10. Three terms

12. One term

14. Two terms

16. One term

18. Two terms

20. $-6^2 = -(6)^2 = -36$

22. $(-4)^2 = 16$

24. $3^2 + (-4)^2 = 9 + 16$
$= 25$

26. $\frac{4^2 - 3^2}{8 - 1} - (2 \cdot 1)^2$
$= \frac{16 - 9}{7} - 4$
$= \frac{7}{7} - 4 = -3$

28. $\frac{3^2 \cdot 2^2}{4 - 1} + \frac{(-3)(2)^3}{6}$
$= \frac{9 \cdot 4}{3} + \frac{(-3)(8)}{6}$
$= \frac{36}{3} + \frac{-24}{6}$
$= 12 + (-4) = 8$

30. $\frac{7^2 - 6^2}{10 + 3} - \frac{8^2 \cdot (-2)}{(-4)^2}$
$= \frac{49 - 36}{13} - \frac{64 \cdot (-2)}{16}$
$= \frac{13}{13} - \frac{-128}{16}$
$= 1 - (-8) = 9$

32. $-5(3)^2(-2) = -5(9)(-2)$
$= 90$

34. $3 - (-2)^2 = 3 - (4)$
$= -1$

36. $[3 + 2(-2)]^2$
$= [3 + (-4)]^2$
$= [-1]^2 = 1$

38. $(3(-2))^2 - 3(3)$
$= (-6)^2 - 9$
$= 36 - 9 = 27$

40. $\dfrac{-3(-2)^2}{6} + 2(3)(-2)$

$= \dfrac{-3(4)}{6} + (-12)$

$= \dfrac{-12}{6} + (-12)$

$= -2 + (-12) = -14$

42. $(3 + (-2))^2 + (3 - (-2))^2$
$= (1)^2 + (5)^2$
$= 1 + 25$
$= 26$

44. $\frac{1}{2}(32)(3)^2 - 12(3)$
$= 16(9) - 36$
$= 144 - 36 = 108$

46. $\dfrac{(32)(12 - 4)^2}{32}$

$= \dfrac{(32)(8)^2}{32} = 64$

48. $\dfrac{(4) - (4)(2)^3}{1 - (2)}$

$= \dfrac{(4) - (4)(8)}{-1}$

$= \dfrac{4 - 32}{-1} = \dfrac{-28}{-1} = 28$

50. $P(3)$
$= 2(3)^3 + (3)^2 - 3(3) + 4$
$= 2(27) + 9 - 9 + 4 = 58;$

$P(-3)$
$= 2(-3)^3 + (-3)^2 - 3(-3) + 4$
$= 2(-27) + 9 + 9 + 4$
$= -32;$

$P(0)$
$= 2(0)^3 + (0)^2 - 3(0) + 4$
$= 4$

52. $D(4) = [3(4) - 1]^2 + 2(4)^2$
$= (12 - 1)^2 + 2(16)$
$= (11)^2 + 32 = 153;$

$D(-4) = [3(-4) - 1]^2 + 2(-4)^2$
$= (-12 -1)^2 + 2(16)$
$= (-13)^2 + 32 = 201;$

$D(0) = [3(0) - 1]^2 + 2(0)^2$
$= (0 - 1)^2 + 0 = (-1)^2 = 1$

54. $P(0) = -3(0)^2 + 1$
$= 0 + 1 = 1;$

$Q(-1) = 2(-1)^2 - (-1) + 1$
$= 2(1) + 1 + 1 = 4$

56. $P(-1) = (-1)^6 - (-1)^5$
$= 1 - (-1) = 2;$

$Q(-1) = (-1)^7 - (-1)^6$
$= -1 - (1) = -2$

58. $P(1) = 1 + 1 = 2,$

$R(-2) = (-2)^2 + (-2) - 1$
$= 1,$

$Q(-1) = (-1)^2 - 1 = 0$

$P(1) + R(-2) \cdot Q(-1)$
$= 2 + (1)(0)$
$= 2 + 0 = 2$

60. $P(3) = 3 + 1 = 4,$

$R(-1) = (-1)^2 + (-1) - 1$
$= -1,$

$Q(1) = 1^2 - 1 = 0;$

$P(3)[R(-1) + Q(1)]$
$= 4[-1 + 0]$
$= 4[-1] = -4$

62. $P(-1) = -1 + 1 = 0;$

$R[P(-1)] = R(0)$

$= 0^2 + 0 - 1$
$= -1$

64. $R(-2) = (-2)^2 + (-2) - 1$
$= 4 + (-2) - 1 = 1;$

$Q[R(-2)] = Q(1)$

$= 1^2 - 1 = 0$

66. Let $x = 2$ and $y = 1$. Then

$$-xy^2 = -2(1)^2 = -2$$

and

$$-(xy)^2 = -(2 \cdot 1)^2 = -4.$$

Therefore, for all x and y, $-xy^2 \neq -(xy)^2$.

68. $(-x)^n = (-x)(-x) \cdots (-x)$, n factors
$= (-1)(x)(-1)(x) \cdots (-1)(x)$, n factors of (-1) and of (x),
$= \big((-1)(-1) \cdots (-1)\big)\big((x)(x) \cdots (x)\big)$ by repeated application of the commutative property and by the associative property of multiplication.

Hence, $(-x)^n = (-1)^n(x)^n$.

If n is odd, $(-1)^n = -1$, and $(-x)^n = (-1)(x)^n = -x^n$.

If n is even, $(-1)^n = 1$, and $(-x)^n = 1(x)^n = x^n$.

EXERCISE 2.2

2. $2b - 7b = -5b$

4. $-8z + 8z = 0$

6. $(2r - 3r) + 6r$
$= -r + 6r$
$= 5r$

8. $5a^2b - 3ab^2 + a^2b$
$= (5a^2b + a^2b) - 3ab^2$
$= 6a^2b - 3ab^2$

10. $(4t^2 + 3t) - (2t^2 + 3t)$
$= 4t^2 + 3t - 2t^2 - 3t$
$= (4t^2 - 2t^2) + (3t - 3t)$
$= 2t^2 + 0 = 2t^2$

12. $(2z^3 - 3z^2) + (3z^2 - 2z^3)$
$= 2z^3 - 3z^2 + 3z^2 - 2z^3$
$= (2z^3 - 2z^3) + (3z^2 - 3z^2)$
$= 0$

14. $(2r^2 - 3r + 3) - (2r^2 - 3r + 1)$
$= 2r^2 - 3r + 3 - 2r^2 + 3r - 1$
$= (2r^2 - 2r^2) + (-3r + 3r) + (3 - 1)$
$= 0 + 0 + 2 = 2$

16. $(3y^2 - 3y + 7) + (2y^2 + y + 6)$
$= 3y^2 - 3y + 7 + 2y^2 + y + 6$
$= (3y^2 + 2y^2) + (-3y + y) + (7 + 6)$
$= 5y^2 + (-2y) + 13$
$= 5y^2 - 2y + 13$

18. $(2z^3 - 3z^2 + 2z) + (4z - 2z^2 - 3z^3)$
$= 2z^3 - 3z^2 + 2z + 4z - 2z^2 - 3z^3$
$= (2z^3 - 3z^3) + (-3z^2 - 2z^2) + (2z + 4z)$
$= -z^3 + (-5z^2) + 6z$
$= -z^3 - 5z^2 + 6z$

20. $(7c^2 - 10c + 8) + (8c + 11) + (-6c^2 - 3c - 2)$
$= 7c^2 - 10c + 8 + 8c + 11 - 6c^2 - 3c - 2$
$= (7c^2 - 6c^2) + (-10c + 8c - 3c) + (8 + 11 - 2)$
$= c^2 - 5c + 17$

22. $(m^2n^2 - 2mn + 7) + (-2m^2n^2 + mn - 3) + (3m^2n^2 - 4mn + 2)$
$= m^2n^2 - 2mn + 7 - 2m^2n^2 + mn - 3 + 3m^2n^2 - 4mn + 2$
$= (m^2n^2 - 2m^2n^2 + 3m^2n^2) + (-2mn + mn - 4mn) + (7 - 3 + 2)$
$= 2m^2n^2 - 5mn + 6$

24. $(-3y^3 + 4y^2 + 6y) + (y^3 - 2y^2 + y + 6) + (4y^3 - 2y^2 - 4y - 1)$
$= (-3y^3 + y^3 + 4y^3) + (4y^2 - 2y^2 - 2y^2) + (6y + y - 4y) + (6 - 1)$
$= 2y^3 + 3y + 5$

26. $(4y^2 - 3y - 7) - (6y^2 - y + 2) = (4y^2 - 3y - 7) + (-6y^2 + y - 2)$
$= 4y^2 - 3y - 7 - 6y^2 + y - 2$
$= (4y^2 - 6y^2) + (-3y + y) + (-7 - 2)$
$= -2y^2 - 2y - 9$

28. $(4s^3 - 3s^2 + 2s - 1) - (s^3 - s^2 + 2s - 1)$
$= (4s^3 - 3s^2 + 2s - 1) + (-s^3 + s^2 - 2s + 1)$
$= (4s^3 - s^3) + (-3s^2 + s^2) + (2s - 2s) + (-1 + 1)$
$= 3s^3 - 2s^2$

30. $(a^3 - 3a^2b - 5ab^2 + 6b^3) - (a^3 + a^2b - 4ab^2 - 5b^3)$
$= (a^3 - 3a^2b - 5ab^2 + 6b^3) + (-a^3 - a^2b + 4ab^2 + 5b^3)$
$= a^3 - 3a^2b - 5ab^2 + 6b^3 - a^3 - a^2b + 4ab^2 + 5b^3$
$= (a^3 - a^3) + (-3a^2b - a^2b) + (-5ab^2 + 4ab^2) + (6b^3 + 5b^3)$
$= -4a^2b - ab^2 + 11b^3$

32. $[(2t^2 - 3t + 5) + (t^2 + t + 2)] - (2t^2 + 3t - 1)$
$= [2t^2 - 3t + 5 + t^2 + t + 2] - (2t^2 + 3t - 1)$
$= [(2t^2 + t^2) + (-3t + t) + (5 + 2)] - (2t^2 + 3t - 1)$
$= 3t^2 + (-2t) + 7 - 2t^2 - 3t + 1$
$= (3t^2 - 2t^2) + \big((-2t) - 3t\big) + (7 + 1)$
$= t^2 + (-5t) + 8$
$= t^2 - 5t + 8$

34. $(2c^2 + 3c + 1) - [(7c^2 + 3c - 2) + (3 - c - 5c^2)]$
$= (2c^2 + 3c + 1) - [(7c^2 - 5c^2) + (3c - c) + (-2 + 3)]$
$= (2c^2 + 3c + 1) - [2c^2 + 2c + 1]$
$= 2c^2 + 3c + 1 - 2c^2 - 2c - 1$
$= (2c^2 - 2c^2) + (3c - 2c) + (1 - 1)$
$= c$

36. $[(y + 2) + (y^2 - 4y + 3)] - (2y^2 - y + 1)$

$= [y^2 + (y - 4y) + (2 + 3)] - (2y^2 - y + 1)$

$= y^2 + (-3y) + 5 - 2y^2 + y - 1$

$= (y^2 - 2y^2) + (-3y + y) + (5 - 1)$

$= -y^2 + (-2y) + 4$

$= -y^2 - 2y + 4$

38. $3a + [2a - (a + 4)] = 3a + [2a - a - 4]$
$= 3a + [a - 4]$
$= 3a + a - 4 = 4a - 4$

40. $5 - [3y + (y - 4) - 1] = 5 - [3y + y - 4 - 1]$
$= 5 - [4y - 5]$
$= 5 - 4y + 5 = 10 - 4y$

42. $-(x - 3) + [2x - (3 + x) - 2] = -x + 3 + [2x - 3 - x - 2]$
$= -x + 3 + [x - 5]$
$= -x + 3 + x - 5 = -2$

44. $[2y^2 - (4 - y)] + [y^2 - (2 + y)]$

$= [2y^2 - 4 + y] + [y^2 - 2 - y]$

$= 2y^2 - 4 + y + y^2 - 2 - y$

$= 3y^2 - 6$

46. $y - (y - [x - (2x + y)] - 2y) = y - (y - [x - 2x - y] - 2y)$
$= y - (y - [-x - y] - 2y)$
$= y - (y + x + y - 2y)$
$= y - x$

48. $[x - (y + x)] - (2x - [3x - (x - y)] + y)$
$= [x - y - x] - (2x - [3x - x + y] + y)$
$= -y - (2x - [2x + y] + y)$
$= -y - (2x - 2x - y + y)$
$= -y - 0 = -y$

50. $-(2y - [2y - 4y + (y - 2)] + 1) + [2y - (4 - y) + 1]$
$= -(2y - [-2y + y - 2] + 1) + [2y - 4 + y + 1]$
$= -(2y - [-y - 2] + 1) + [3y - 3]$
$= -(2y + y + 2 + 1) + [3y - 3]$
$= -(3y + 3) + [3y - 3]$
$= -3y - 3 + 3y - 3$
$= -6$

52. $[2x + (x - y) + 2y] - (3y - [x + (y - x) + y] + 2x)$
$= [2x + x - y + 2y] - (3y - [x + y - x + y] + 2x)$
$= [3x + y] - (3y - [2y] + 2x)$
$= 3x + y - (y + 2x)$
$= 3x + y - y - 2x$
$= x$

54. $P(x) - Q(x) + R(x) = x - 1 - (x^2 + 1) + (x^2 - x + 1)$
$= x - 1 - x^2 - 1 + x^2 - x + 1$
$= -1$

56. $Q(x) - [R(x) + P(x)] = x^2 + 1 - [(x^2 - x + 1) + (x - 1)]$
$= x^2 + 1 - [x^2]$
$= x^2 + 1 - x^2$
$= 1$

58. Let $x = 1$ and $y = 2$. Then

$$-(x - y) = -(1 - 2) = 1$$

and

$$-x - y = -1 - 2 = -3$$

Therefore, $-(x - y)$ is not equivalent to $-x - y$.

EXERCISE 2.3

2. $(4c^3)(3c) = 4 \cdot 3c^{3+1}$
$= 12c^4$

4. $(-6r^2s^2)(5rs^3)$
$= -6 \cdot 5r^{2+1}s^{2+3}$
$= -30r^3s^5$

6. $(-8abc)(-b^2c^3)$
$= (-8abc)(-1b^2c^3)$
$= -8 \cdot (-1)ab^{1+2}c^{1+3}$
$= 8ab^3c^4$

8. $-5(ab^3)(-3a^2bc)$
$= -5(-3)a^{1+2}b^{3+1}c$
$= 15a^3b^4c$

10. $(-5mn)(2m^2n)(-n^3)$
$= (-5mn)(2m^2n)(-1 \cdot n^3)$
$= -5 \cdot 2 \cdot (-1)m^{1+2}n^{1+1+3}$
$= 10m^3n^5$

12. $(-3xy)(2xz^4)(3x^3y^2z)$
$= -3 \cdot 2 \cdot 3x^{1+1+3}y^{1+2}z^{4+1}$
$= -18x^5y^3z^5$

14. $-a^2(ab^2)(2a)(-3b^2)$
$= -1 \cdot 2 \cdot (-3)a^{2+1+1}b^{2+2}$
$= 6a^4b^4$

16. $(-2y^2)(y^2)(y)$
$= -2 \cdot y^{2+2+1}$
$= -2y^5$

18. If any one factor in a product equals zero, then the product equals zero. Therefore, $(-t)(2t^2)(-t)(0) = 0$.

20. $\dfrac{12a^4b^2}{-4a^2b^2} = -3a^{4-2} \cdot \dfrac{b^2}{b^2}$
$= -3^2 \cdot 1 = -3a^2$

22. $\dfrac{-22a^2bc^3}{-11ac^2} = 2a^{2-1}bc^{3-2}$
$= 2abc$

24. $\dfrac{100m^2n^3}{5m^2n^3} = 20 \cdot \dfrac{m^2}{m^2} \cdot \dfrac{n^3}{n^3}$
$= 20 \cdot 1 \cdot 1 = 20$

26. $\dfrac{-x^4y^8z^6}{xy^7z^5} = -x^{4-1}y^{8-7}z^{6-5}$
$= -x^3yz$

28. $\dfrac{(t + 3)^5}{(t + 3)^3} = (t + 3)^{5-3}$
$= (t + 3)^2$

30. $\dfrac{34a^6b^2c^4}{-17abc^3} = -2a^{6-1}b^{2-1}c^{4-3}$
$= -2a^5bc$

32. $\dfrac{-15xy^3z^4}{-3y^2z^4} = 5xy^{3-2} \cdot \dfrac{z^4}{z^4}$
$= 5xy \cdot 1$
$= 5xy$

34. $\dfrac{16(x - 3)^2}{-8(x - 3)^2} = -2 \cdot \dfrac{(x - 3)^2}{(x - 3)^2}$
$= -2 \cdot 1$
$= -2$

36. $b^{-n} \cdot b^{2n+1} = b^{-n+2n+1}$
$= b^{n+1}$

38. $y^{2n+6} \cdot y^{4-n} = y^{2n+6+4-n}$
$= y^{n+10}$

40. $b^{n+2} \cdot b^{2n-1} = b^{n+2+2n-1}$
$= b^{3n+1}$

42. $\dfrac{b^{7n}}{b^n} = b^{7n-n}$
$= b^{6n}$

44. $\dfrac{x^{3n+4}}{x^{2n-1}} = x^{3n+4-(2n-1)}$
$= x^{3n+4-2n+1}$
$= x^{n+5}$

46. $\dfrac{y^{n^2+2n-2}}{y^{n^2-2}} = y^{n^2+2n-2-(n^2-2)}$
$= y^{n^2+2n-2-n^2+2}$
$= y^{2n}$

48. $\dfrac{x^{n-3}b^{n+5}}{x^{n-4}b^{n+1}} = x^{n-3-(n-4)}b^{n+5-(n+1)}$
$= x^{n-3-n+4}b^{n+5-n-1}$
$= xb^4$

50. $$(xy)^n = \overbrace{xy \cdot xy \cdot xy \cdot \;\cdots\; \cdot xy}^{n \text{ factors}}$$

Using the commutative law of multiplication repeatedly,

$$(xy)^n = \overbrace{x \cdot x \cdot x \cdot \;\cdots\; \cdot x}^{n \text{ factors}} \cdot \overbrace{y \cdot y \cdot y \cdot \;\cdots\; \cdot y}^{n \text{ factors}}$$
$$= x^n y^n$$

EXERCISE 2.4

2. $-3y(2x + y)$
$= -3y(2x) + (-3y)(y)$
$= -6xy + (-3y^2)$
$= -6xy - 3y^2$

4. $4y(2y^2 - y - 3)$
$= 4y(2y^2) - 4y(y) - 4y(3)$
$= 8y^3 - 4y^2 - 12y$

6. $-2(t^3 - 3t^2 + 2t - 1) = -2t^3 + 6t^2 - 4t + 2$

8. $(2x - 1)^2 = (2x - 1)(2x - 1)$
$= (2x)(2x) + [(-1)(2x) + 2x(-1)] + (-1)(-1)$
$= 4x^2 + [(-2x) + (-2x)] + 1$
$= 4x^2 - 4x + 1$

10. $(n + 4)^2 = (n + 4)(n + 4)$
$= n \cdot n + [4n + 4n] + 4 \cdot 4 = n^2 + 8n + 16$

12. $(z - 3)(z + 5) = z \cdot z + [-3z + 5z] + (-3 \cdot 5)$
$= z^2 + 2z - 15$

14. $(r - 1)(r - 6) = r \cdot r + [-1 \cdot r + (-6 \cdot r)] + (-1)(-6)$
$= r^2 - 7r + 6$

16. $(y - 2)(y + 3) = (y)(y) + [(-2)(y) + (y)(3)] + (-2)(3)$
$= y^2 + [-2y + 3y] + (-6) = y^2 + y - 6$

18. $(z - 3)(z - 5)$
$= z^2 - 8z + 15$

20. $(3t - 1)(2t + 1)$
$= 6t^2 + t - 1$

22. $(2z - 1)(3z + 5)$
$= 6z^2 + 7z - 5$

24. $(3t - 4s)(3t + 4s)$
$= 9t^2 - 16s^2$

For Exercises 26-36, only Method 1 will be shown.

26. $(t + 4)(t^2 - t - 1) = t(t^2 - t - 1) + 4(t^2 - t - 1)$
$= t^3 - t^2 - t + 4t^2 - 4t - 4$
$= t^3 + 3t^2 - 5t - 4$

28. $(x - 7)(x^2 - 3x + 1) = x(x^2 - 3x + 1) - 7(x^2 - 3x + 1)$
$= x^3 - 3x^2 + x - 7x^2 + 21x - 7$
$= x^3 - 10x^2 + 22x - 7$

30. $(y + 2)(y - 2)(y + 4) = [(y + 2)(y - 2)](y + 4)$
$= [y^2 - 4](y + 4)$
$= y^2(y + 4) - 4(y + 4)$
$= y^3 + 4y^2 - 4y - 16$

32. $(z - 5)(z + 6)(z - 1) = [(z - 5)(z + 6)](z - 1)$
$= [z^2 + z - 30](z - 1)$
$= z^2(z - 1) + z(z - 1) - 30(z - 1)$
$= z^3 - z^2 + z^2 - z - 30z + 30$
$= z^3 - 31z + 30$

34. $(3x - 2)(4x^2 + x - 2) = 3x(4x^2 + x - 2) - 2(4x^2 + x - 2)$
$= 12x^3 + 3x^2 - 6x - 8x^2 - 2x + 4$
$= 12x^3 - 5x^2 - 8x + 4$

36. $(b^2 - 3b + 5)(2b^2 - b + 1)$
$= b^2(2b^2 - b + 1) - 3b(2b^2 - b + 1) + 5(2b^2 - b + 1)$
$= 2b^4 - b^3 + b^2 - 6b^3 + 3b^2 - 3b + 10b^2 - 5b + 5$
$= 2b^4 - 7b^3 + 14b^2 - 8b + 5$

38. $3[2a - (a + 1) + 3] = 3[2a - a - 1 + 3]$
$= 3[a + 2]$
$= 3a + 6$

40. $-2a[3a + (a - 3) - (2a + 1)] = -2a[3a + a - 3 - 2a - 1]$
$= -2a[2a - 4] = -4a^2 + 8a$

42. $-[(a + 1) - 2(3a - 1) + 4] = -[a + 1 - 6a + 2 + 4]$
$= -[-5a + 7]$
$= 5a - 7$

44. $-4(4 - [3 - 2(a - 1) + a] + a) = -4(4 - [3 - 2a + 2 + a] + a]$
$= -4(4 - [-a + 5] + a)$
$= -4(4 + a - 5 + a)$
$= -4(2a - 1) = -8a + 4$

46. $x(4 - 2[3 - 4(x + 1)] - x) = x(4 - 2[3 - 4x - 4] - x)$
$= x(4 - 2[-4x - 1] - x)$
$= x(4 + 8x + 2 - x)$
$= x(7x + 6)$
$= 7x^2 + 6x$

48. $3t^n(2t^n + 3)$
$= 3t^n(2t^n) + 3t^n(3)$
$= 6t^{n+n} + 9t^n$
$= 6t^{2n} + 9t^n$

50. $b^{n-1}(b + b^n)$
$= b^{n-1}(b) + b^{n-1}(b^n)$
$= b^{n-1+1} + b^{n-1+n}$
$= b^n + b^{2n-1}$

52. $b^{2n+2}(b^{n-1} + b^n) = b^{2n+2}(b^{n-1}) + b^{2n+2}(b^n)$
$= b^{2n+2+n-1} + b^{2n+2+n}$
$= b^{3n+1} + b^{3n+2}$

54. $(a^n - 3)(a^n + 2)$
$= a^{2n} - 3a^n + 2a^n - 6$
$= a^{2n} - a^n - 6$

56. $(a^{3n} - 3)(a^{3n} + 3)$
$= a^{6n} - 9$

58. $(a^{2n} - 2b^n)(a^{3n} + b^{2n}) = a^{2n}(a^{3n} + b^{2n}) - 2b^n(a^{3n} + b^{2n})$
$= a^{5n} + a^{2n}b^{2n} - 2a^{3n}b^n - 2b^{3n}$

60. $(x + a)^2 = (x + a)(x + a) = x \quad + 2ax + a^2$

62. $(ax + by)(cx + dy) = ax(cx + dy) + by(cx + dy)$
$= acx^2 + adxy + bcxy + bdy^2;$
$= acx^2 + (ad + bc)xy + bdy^2$

Therefore, $(ax + by)(cx + dy) = acx^2 + (ad + bc)xy + bdy^2$.

64. $(x - a)(x^2 + ax + a^2) = x(x^2 + ax + a^2) - a(x^2 + ax + a^2)$
$= x^3 + ax^2 + a^2x - ax^2 - a^2x - a^3$
$= x^3 - a^3$

66. Let $x = 2$ and $y = 1$. Then

$$(x - y)^2 = (2 - 1)^2 = 1$$

and

$$x^2 - y^2 = 2^2 - 1^2 = 3.$$

Therefore, $(x - y)^2$ is not equivalent to $x^2 - y^2$.

EXERCISE 2.5

2. $2 \cdot 13$

4. $2 \cdot 3 \cdot 3$

6. $-1 \cdot 2 \cdot 2 \cdot 2 \cdot 2$

8. Prime

10. $5 \cdot 13$

12. $3x - 9 = 3(? - ?)$
$= 3(x - 3)$

14. $3xy(? + ?)$
$= 3xy(x + 2)$

16. $x(? - ? + ?)$
$= x(x^2 - x + 1)$

18. $3rs(? + ? - ?)$
$= 3rs(5r + 6s - 1)$

20. $3n^2(? - ? + ?)$
$= 3n^2(n^2 - 2n + 4)$

22. $x(? - ? + ?)$
$= x(2xy^2 - 3y + 5x)$

24. $xz(? + ? - ?)$
$= xz(xy^2z + 2y - 1)$

26. $3x(? - ? + ?)$
$= 3x(2xy - 3y^2 + 4)$

28. $7xy(? + ? - ?)$
$= 7xy(2 + 3xy - 4z)$

30. $(a - 2)(? + ?)$
$= (a - 2)(b + a)$

32. $(y - 2)(? + ?)$
$= (y - 2)(y - 3x)$

34. $(2a - b)(? + ?)$
$= (2a - b)(3x + 4y)$

36. $-(-3m + 2n)$
$= -(2n - 3m)$

38. $-(-r^2 + s^2t^2)$
$= -(s^2t^2 - r^2)$

40. $-3(? + ?)$
$= -3(2x + 3)$

42. $?(a - b)$
$= -a(a - b)$

44. $-(? + ?)$
$= -(-x^2 + 3)$
$= -(3 - x^2)$

46. $-(? + ? + ?)$
$= -(-3x - 3y + 2z)$

48. $x^{2n} \cdot x^{2n} - x^{2n} \cdot 1 = x^{2n}(x^{2n} - 1)$
$= x^{2n}(x^n + 1)(x^n - 1)$

50. $y^{2n} \cdot y^{2n} + y^{2n} \cdot y^n + y^{2n} \cdot 1 = y^{2n}(y^{2n} + y^n + 1)$

52. $x^n \cdot x^2 + x^n \cdot x + x^n \cdot 1 = x^n(x^2 + x + 1)$

54. $(-x^{2n})x^{3n} - (-x^{2n}) \cdot 1 = -x^{2n}(x^{3n} - 1)$

56. $(-y^2)y^a - (-y^2) \cdot 1 = -y^2(y^a - 1)$

EXERCISE 2.6

2. $(x + 2)(x + 5)$

4. $(a - 5)(a + 3)$

6. $(x - 4y)(x - 5y)$

8. $(x + 6)(x - 6)$

10. $(3 + a)(3 - a)$

12. $(a^2b + 2)(a^2b - 2)$

14. $(y^2 + 5)(y^2 - 5)$

16. $(x + 2y)(x - 2y)$

18. $(x)^2 - (9y)^2$
$= (x + 9y)(x - 9y)$

20. $(3x - 1)(x - 2)$

22. $(3x - 1)(2x - 1)$

24. $(4a - 1)(a - 1)$

26. $(2x - 3)(5x + 6)$

28. $(3x + 5a)(3x - 2a)$

30. $(2x + 3y)(2x - 3y)$

32. $(2y + 1)(2y + 1)$
$= (2y + 1)^2$

34. $(8xy)^2 - 1^2$
$= (8xy + 1)(8xy - 1)$

36. $(2xy + 3)(2xy + 3)$
$= (2xy + 3)^2$

38. $2(x^2 + 3x - 10)$
$= 2(x + 5)(x - 2)$

40. $a(2a^2 + 15a + 7)$
$= a(a + 7)(2a + 1)$

42. $5(4a^2 + 12ab + 9b^2)$
$= 5(2a + 3b)(2a + 3b)$
$= 5(2a + 3b)^2$

44. $x^2(1 - 4y^2)$
$= x^2(1 + 2y)(1 - 2y)$

46. $x^2(1 - 2x + x^2) = x^2(1 - x)^2$, or

$x^2(1 - 2x + x^2) = x^2(x^2 - 2x + 1)$
$= x^2(x - 1)^2$

48. $xy(x^2 - y^2)$
$= xy(x + y)(x - y)$

50. $(a^2 + 2)(a^2 + 3)$

52. $(4x^2 + 1)(x^2 - 3)$

54. $(x^2 - 9)(x^2 + 3)$
$= (x + 3)(x - 3)(x^2 + 3)$

56. $(y^2 - 4)(y^2 - 9) = (y + 2)(y - 2)(y + 3)(y - 3)$

58. $(x^2 - 4)(3x^2 + 1) = (x + 2)(x - 2)(3x^2 + 1)$

60. $(x^2 - 9a^2)(4x^2 + 3a^2) = (x + 3a)(x - 3a)(4x^2 + 3a^2)$

62. $16 - y^{4n}$
$= (4)^2 - (y^{2n})^2$
$= (4 - y^{2n})(4 + y^{2n})$
$= [(2)^2 - (y^n)^2](4 + y^{2n})$
$= (2 - y^n)(2 + y^n)(4 + y^{2n})$

64. $x^{4n} - 2x^{2n} + 1$
$= (x^{2n})^2 - 2(x^{2n}) + 1$
$= (x^{2n} - 1)(x^{2n} - 1)$
$= (x^{2n} - 1)^2$
$= [(x^n)^2 - (1)^2]^2$
$= [(x^n - 1)(x^n + 1)]^2$
$= (x^n - 1)^2(x^n + 1)^2$

66. $6(y^{2n} + 5y^n - 150) = 6(y^n - 10)(y^n + 15)$

EXERCISE 2.7

2. $a(5 + b) + b(5 + b)$
$= (5 + b)(a + b)$

4. $a(1 + b) + b(1 + b)$
$= (1 + b)(a + b)$

6. $(x - y)x^2 + (x - y)y$
$= (x - y)(x^2 + y)$

8. $1 \cdot (1 - x) - y(1 - x)$
$= (1 - x)(1 - y)$

10. $5z(x - y) - 1 \cdot (x - y)$
$= (x - y)(5z - 1)$

12. $2x^2(3x - 2) + 1 \cdot (3x - 2)$
$= (3x - 2)(2x^2 + 1)$

14. $a(2a + 3) - b(2a + 3)$
$= (2a + 3)(a - b)$

16. $a(2b^2 + 5) - 4(2b^2 + 5)$
$= (2b^2 + 5)(a - 4)$

18. $4(3 - y^3) - x^2(3 - y^3)$
$= (3 - y^3)(4 - x^2)$
$= (3 - y^3)(2 + x)(2 - x)$

20. $y^3 + 3^3$
$= (y + 3)(y^2 - 3y + 9)$

22. $y^3 - (3x)^3 = (y - 3x)[y^2 + y(3x) + (3x)^2]$
$= (y - 3x)(y^2 + 3xy + 9x^2)$

24. $(3a)^3 + b^3 = (3a + b)[(3a)^2 - (3a)b + b^2]$
$= (3a + b)(9a^2 - 3ab + b^2)$

26. $2^3 + (xy)^3 = (2 + xy)[2^2 - 2(xy) + (xy)^2]$
$= (2 + xy)(4 - 2xy + x^2y^2)$

28. $a^3 - (5b)^3 = (a - 5b)[a^2 + a(5b) + (5b)^2]$
$= (a - 5b)(a^2 + 5ab + 25b^2)$

30. $[(x + y) - z][(x + y)^2 + (x + y)z + z^2]$
$= (x + y - z)(x^2 + 2xy + y^2 + xz + yz + z^2)$

32. $(x^2)^3 + (x - 2y)^3$
$= [(x^2) + (x - 2y)][(x^2)^2 - (x^2)(x - 2y) + (x - 2y)^2]$
$= (x^2 + x - 2y)(x^4 - x^3 + 2x^2y + x^2 - 4xy + 4y^2)$

34. $[(2y - 1) + (y - 1)][(2y - 1)^2 - (2y - 1)(y - 1) + (y - 1)^2]$
$= (2y - 1 + y - 1)[4y^2 - 4y + 1 - (2y^2 - 3y + 1) + y^2 - 2y + 1]$
$= (3y - 2)(4y^2 - 4y + 1 - 2y^2 + 3y - 1 + y^2 - 2y + 1)$
$= (3y - 2)(3y^2 - 3y + 1)$

36. $[(x + y) - (x - y)][(x + y)^2 + (x + y)(x - y) + (x - y)^2]$
$= (x + y - x + y)(x^2 + 2xy + y^2 + x^2 - y^2 + x^2 - 2xy + y^2)$
$= 2y(3x^2 + y^2)$

38. $a^2 - b^2 - c^2 + 2bc = a^2 - b^2 + 2bc - c^2$
$= a^2 - (b^2 - 2bc + c^2)$
$= (a)^2 - (b - c)^2$
$= [(a) - (b - c)][(a) + (b - c)]$
$= (a - b + c)(a + b - c)$

40. $x^4 - 8x^2y^2 + 4y^4 = x^4 - 8x^2y^2 + 4y^4 + 4x^2y^2 - 4x^2y^2$

$= (x^4 - 4x^2y^2 + 4y^4) - 4x^2y^2$

$= (x^2 - 2y^2)^2 - (2xy)^2$

$= (x^2 - 2y^2 - 2xy)(x^2 - 2y^2 + 2xy)$

42. $4a^4 - 5a^2b^2 + b^4 = 4a^4 - 5a^2b^2 + b^4 + a^2b^2 - a^2b^2$

$= (4a^4 - 4a^2b^2 + b^4) - a^2b^2$

$= (2a^2 - b^2)^2 - (ab)^2$

$= (2a^2 - b^2 - ab)(2a^2 - b^2 + ab)$

3

FRACTIONS

EXERCISE 3.1

2. $\frac{-3}{4}$

4. $\frac{-6}{7}$

6. $\frac{-3}{5}$

8. $\frac{1}{2}$

10. $\frac{-2y}{x}$ $(x \neq 0)$

12. $\frac{x^3}{y^2}$ $(y \neq 0)$

14. $\frac{-(x + 2)}{x}$ or $\frac{-x - 2}{x}$ $(x \neq 0)$

16. $\frac{-(2y - x)}{2x}$ or $\frac{x - 2y}{2x}$ $(x \neq 0)$

18. $\frac{-3}{-(x - 2)} = \frac{3}{x - 2}$

20. $\frac{-6}{-(y - x)} = \frac{6}{y - x}$

22. $\frac{2x - 5}{-(y - 3)} = \frac{-(2x - 5)}{y - 3}$ or $\frac{5 - 2x}{y - 3}$

24. $\frac{x - 3}{-(x - y)} = \frac{-(x - 3)}{x - y}$ or $\frac{3 - x}{x - y}$

26. $-\frac{x}{-(2y - x)} = \frac{x}{2y - x}$

28 $\frac{-a}{-(3a - 2b)} = \frac{a}{3a - 2b}$

30. Let $x = 2$ and $y = 1$. Then

$$-\frac{2x - y}{3} = -\frac{2(2) - 1}{3} = -1 \quad \text{and} \quad \frac{-2x - y}{3} = \frac{-2(2) - 1}{3} = \frac{-5}{3}.$$

Therefore, $-\frac{2x - y}{3}$ is not equivalent to $\frac{-2x - y}{3}$.

EXERCISE 3.2

2. $\dfrac{xz \cdot x^3y^2}{1 \cdot x^3y^2} = xz$

4. $\dfrac{1 \cdot 3r}{3rt^3 \cdot 3r} = \dfrac{1}{3rt^3}$

6. $\dfrac{4(x - 1)}{(x - 1)} = 4$

8. $\dfrac{3(2x + 3)}{5(2x + 3)} = \dfrac{3}{5}$

10. $\dfrac{(x - 4)(x + 4)}{x + 4} = x - 4$

12. $\dfrac{4(x^2 - 4)}{x + 1} = \dfrac{4(x - 1)(x + 1)}{x + 1}$
$= 4(x - 1)$

14. $\dfrac{5(y^2 - 4)}{y - 2}$
$= \dfrac{5(y + 2)(y - 2)}{y - 2}$
$= 5(y + 2)$

16. $\dfrac{-1(y - 2)}{(y + 2)(y - 2)} = \dfrac{-1}{y + 2}$

18. $\dfrac{2x - y}{-(4x^2 - y^2)} = \dfrac{2x - y}{-(2x - y)(2x + y)} = \dfrac{1}{-(2x + y)} = \dfrac{-1}{2x + y}$

20. $\dfrac{2(y - 4)}{2 \cdot 4} = \dfrac{y - 4}{4}$

22. $\dfrac{5(y - 2)}{5 \cdot 1} = y - 2$

24. $\dfrac{b(x^2 + 1)}{b(x + 1)} = \dfrac{x^2 + 1}{x + 1}$

26. $\dfrac{-3x(-x^2 + 2x - 1)}{-3x}$
$= -x^2 + 2x - 1$

28. $\dfrac{(x + 2)(x + 3)}{(x + 3)} = x + 2$

30. $\dfrac{(y + 1)(y - 3)}{(y + 3)(y - 3)} = \dfrac{y + 1}{y + 3}$

32. $\dfrac{(3x + 1)(2x - 1)}{(x + 5)(2x - 1)}$
$= \dfrac{3x + 1}{x + 5}$

34. $\dfrac{(2x + 3y)(2x - 3y)}{(x + 2y)(2x - 3y)}$
$= \dfrac{2x + 3y}{x + 2y}$

36. $\dfrac{2x(x + 3) - y(x + 3)}{(2x - y)} = \dfrac{(x + 3)(2x - y)}{(2x - y)} = x + 3$

38. $\dfrac{3 \cdot 2}{4 \cdot 2} = \dfrac{6}{8}$

40. $\dfrac{-12 \cdot 4}{5 \cdot 4} = \dfrac{-48}{20}$

42. $\dfrac{6 \cdot 7}{1 \cdot 7} = \dfrac{42}{7}$

44. $\dfrac{5 \cdot 7}{3y \cdot 7} = \dfrac{35}{21y}$

46. $\frac{-a(ab)}{b(ab)} = \frac{-a^2b}{ab^2}$

48. $\frac{x(xy^3)}{1(xy^3)} = \frac{x^2y^3}{xy^3}$

50. $\frac{2(x - y)}{5(x - y)} = \frac{2x - 2y}{5(x - y)}$

52. $\frac{(a - 2)[2(a - 3)]}{3[2(a - 3)]} = \frac{2(a - 2)(a - 3)}{6(a - 3)} = \frac{2a^2 - 10a + 12}{6(a - 3)}$

54. $\frac{5}{(2a + b)} = \frac{?}{(2a + b)(2a - b)}$; $\frac{5(2a - b)}{(2a + b)(2a - b)} = \frac{10a - 5b}{4a^2 - b^2}$

56. $\frac{5x}{(y + 3)} = \frac{?}{(y + 3)(y - 2)}$; $\frac{5x(y - 2)}{(y + 3)(y - 2)} = \frac{5xy - 10x}{y^2 + y - 6}$

58. $\frac{-3}{(a + 2)} = \frac{?}{(a + 2)(a + 1)}$; $\frac{-3(a + 1)}{(a + 2)(a + 1)} = \frac{-3a - 3}{a^2 + 3a + 2}$

60. $y^2 - x^2 = (-1)(x^2 - y^2) = (-1)(x - y)(x + y)$

$\frac{7}{(x - y)} = \frac{?}{(-1)(x - y)(x + y)}$;

$\frac{7[(-1)(x + y)]}{(x - y)[(-1)(x + y)]} = \frac{-7(x + y)}{(-1)(x - y)(x + y)} = \frac{-7x - 7y}{y^2 - x^2}$

62. $\frac{y}{3 - 2y} = \frac{-y}{2y - 3} = \frac{?}{(2y - 3)(y + 1)}$;

$\frac{-y(y + 1)}{(2y - 3)(y + 1)} = \frac{-y^2 - y}{2y^2 - y - 3}$

64. $\frac{2}{2x - 3y} = \frac{?}{(2x - 3y)(4x^2 + 6xy + 9y^2)}$;

$\frac{2(4x^2 + 6xy + 9y^2)}{(2x - 3y)(4x^2 + 6xy + 9y^2)} = \frac{8x^2 + 12xy + 18y^2}{8x^3 - 27y^3}$

66. $6bx - 3ax + 4by - 2ay = 3x(2b - a) + 2y(2b - a)$
$= (2b - a)(3x + 2y)$;

$\frac{5a - 3b}{3x + 2y} = \frac{?}{(2b - a)(3x + 2y)}$;

$\frac{(5a - 3b)(2b - a)}{(3x + 2y)(2b - a)} = \frac{10ab - 6b^2 - 5a^2 + 3ab}{6bx - 3ax + 4by - 2ay}$

68. $4ax - 3bx - 32ay + 24by = x(4a - 3b) - 8y(4a - 3b)$
$= (4a - 3b)(x - 8y);$

$$\frac{4a + 3b}{x - 8y} = \frac{?}{(x - 8y)(4a - 3b)};$$

$$\frac{(4a + 3b)(4a - 3b)}{(x - 8y)(4a - 3b)} = \frac{16a^2 - 9b^2}{4ax - 3bx - 32ay + 24by}$$

70. Let $x = 2$ and $y = 1$. Then

$$\frac{4x - y}{4} = \frac{4(2) - 1}{4} = \frac{7}{4}$$

and

$$x - y = 2 - 1 = 1$$

Therefore, $\frac{4x - y}{4}$ is not equivalent to $x - y$.

EXERCISE 3.3

2.
$$\begin{array}{r|l} & n + 7 \\ \hline 2n - 1 & 2n^2 + 13n - 7 \\ & 2n^2 - n \\ \hline & 14n - 7 \\ & 14n - 7 \\ \hline & 0 \end{array}$$

$$\frac{2n^2 + 13n - 7}{2n - 1} = n + 7$$

4.
$$\begin{array}{r|l} & x - 4 \\ \hline 2x + 5 & 2x^2 - 3x - 15 \\ & 2x^2 + 5x \\ \hline & -8x - 15 \\ & -8x - 20 \\ \hline & 5 \end{array}$$

$$\frac{2x^2 - 3x - 15}{2x + 5}$$
$$= x - 4 + \frac{5}{2x + 5}$$

6.
$$\begin{array}{r|l} & 2x^2 - 5x + 3 \\ \hline x + 1 & 2x^3 - 3x^2 - 2x + 4 \\ & 2x^3 + 2x^2 \\ \hline & -5x^2 - 2x \\ & -5x^2 - 5x \\ \hline & 3x + 4 \\ & 3x + 3 \\ \hline & 1 \end{array}$$

$$\frac{2x^3 - 3x^2 - 2x + 4}{x + 1} = 2x^2 - 5x + 3 + \frac{1}{x + 1}$$

8.
$$\begin{array}{r|l} & 2b^3 + 8b^2 + 34b + 136 \\ \hline b - 4 & 2b^4 + 0b^3 + 2b^2 + 0b + 3 \\ & \underline{2b^4 - 8b^3} \\ & \quad 8b^3 + 2b^2 \\ & \quad \underline{8b^3 - 32b^2} \\ & \qquad 34b^2 + 0b \\ & \qquad \underline{34b^2 - 136b} \\ & \qquad\quad 136b + 3 \\ & \qquad\quad \underline{136b - 544} \\ & \qquad\qquad 547 \end{array}$$

$$\frac{2b^4 + 2b^2 + 3}{b - 4} = 2b^3 + 8b^2 + 34b + 136 + \frac{547}{b - 4}$$

10.
$$\begin{array}{r|l} & y^4 + y^3 + y^2 + y + 1 \\ \hline y - 1 & y^5 + 0y^4 + 0y^3 + 0y^2 + 0y + 1 \\ & \underline{y^5 - y^4} \\ & \quad y^4 + 0y^3 \\ & \quad \underline{y^4 - y^3} \\ & \qquad y^3 + 0y^2 \\ & \qquad \underline{y^3 - y^2} \\ & \qquad\quad y^2 + 0y \\ & \qquad\quad \underline{y^2 - y} \\ & \qquad\qquad y + 1 \\ & \qquad\qquad \underline{y - 1} \\ & \qquad\qquad\quad 2 \end{array}$$

$$\frac{y^5 + 1}{y - 1} = y^4 + y^3 + y^2 + y + 1 + \frac{2}{y - 1}$$

12.
$$\begin{array}{r|l} & 5t^3 - 9t^2 + 2t - 3 \\ \hline 2t + 3 & 10t^4 - 3t^3 - 23t^2 + 0t + 7 \\ & \underline{10t^4 + 15t^3} \\ & \quad - 18t^3 - 23t^2 \\ & \quad \underline{- 18t^3 - 27t^2} \\ & \qquad 4t^2 + 0t \\ & \qquad \underline{4t^2 + 6t} \\ & \qquad\quad - 6t + 7 \\ & \qquad\quad \underline{- 6t - 9} \\ & \qquad\qquad 16 \end{array}$$

$$\frac{10t^4 - 3t^3 - 23t^2 + 7}{2t + 3} = 5t^3 - 9t^2 + 2t - 3 + \frac{16}{2t + 3}$$

14. $$\frac{15t^3}{3t^2} - \frac{12t^2}{3t^2} + \frac{5t}{3t^2} = \frac{5t \cdot 3t^2}{1 \cdot 3t^2} - \frac{4 \cdot 3t^2}{1 \cdot 3t^2} + \frac{5 \cdot t}{3t \cdot t}$$

$$= 5t - 4 + \frac{5}{3t}$$

16. $$\frac{21n^4}{7n^2} + \frac{14n^2}{7n^2} - \frac{7}{7n^2} = \frac{3n^2 \cdot 7n^2}{1 \cdot 7n^2} + \frac{2 \cdot 7n^2}{1 \cdot 7n^2} - \frac{1 \cdot 7}{n^2 \cdot 7}$$

$$= 3n^2 + 2 - \frac{1}{n^2}$$

18. $$\frac{12x^3}{4x} - \frac{8x^2}{4x} + \frac{3x}{4x} = \frac{3x^2 \cdot 4x}{1 \cdot 4x} - \frac{2x \cdot 4x}{1 \cdot 4x} + \frac{3 \cdot x}{4 \cdot x} = 3x^2 - 2x + \frac{3}{4}$$

20. $$\frac{9a^2b^2}{ab^2} + \frac{3ab^2}{ab^2} + \frac{4a^2b}{ab^2} = \frac{9a \cdot ab^2}{1 \cdot ab^2} + \frac{3 \cdot ab^2}{1 \cdot ab^2} + \frac{4a \cdot ab}{b \cdot ab}$$

$$= 9a + 3 + \frac{4a}{b}$$

22. $$\frac{36t^5}{-12t^2} + \frac{24t^3}{-12t^2} - \frac{12t}{-12t^2} = \frac{3t^3 \cdot 12t^2}{-1 \cdot 12t^2} + \frac{2t \cdot 12t^2}{-1 \cdot 12t^2} - \frac{1 \cdot 12t}{-t \cdot 12t}$$

$$= -3t^3 + (-2t) - \left(-\frac{1}{t}\right) = -3t^3 - 2t + \frac{1}{t}$$

24. $$\frac{15s^{10}}{3s^2} - \frac{21s^5}{3s^2} + \frac{6}{3s^2} = \frac{5s^8 \cdot 3s^2}{1 \cdot 3s^2} - \frac{7s^3 \cdot 3s^2}{1 \cdot 3s^2} + \frac{2 \cdot 3}{s^2 \cdot 3}$$

$$= 5s^8 - 7s^3 + \frac{2}{s^2}$$

26.
$$\begin{array}{r|l} & 2y + 7 \\ \hline y^2 - y - 3 & 2y^3 + 5y^2 - 3y + 2 \\ & \underline{2y^3 - 2y^2 - 6y} \\ & \qquad 7y^2 + 3y + 2 \\ & \qquad \underline{7y^2 - 7y - 21} \\ & \qquad\qquad 10y + 23 \end{array}$$

$$\frac{2y^3 + 5y^2 - 3y + 2}{y^2 - y - 3} = 2y + 7 + \frac{10y + 23}{y^2 - y - 3}$$

28.
$$
\begin{array}{r|l}
 & 2b^2 - 2b + 5 \\ \hline
b^2 + b - 3 & 2b^4 + 0b^3 - 3b^2 + b + 2 \\
 & \underline{2b^4 + 2b^3 - 6b^2} \\
 & \quad - 2b^3 + 3b^2 + b \\
 & \quad \underline{- 2b^3 - 2b^2 + 6b} \\
 & \qquad 5b^2 - 5b + 2 \\
 & \qquad \underline{5b^2 + 5b - 15} \\
 & \qquad\quad -10b + 17
\end{array}
$$

$$\frac{2b^4 - 3b^2 + b + 2}{b^2 + b - 3} = 2b^2 - 2b + 5 + \frac{-10b + 17}{b^2 + b - 3}$$

30.
$$
\begin{array}{r|l}
 & r + 1 \\ \hline
r^3 + 0r^2 + 2r + 3 & r^4 + r^3 - 2r^2 + r + 5 \\
 & \underline{r^4 + 0r^3 + 2r^2 + 3r} \\
 & \quad r^3 - 4r^2 - 2r + 5 \\
 & \quad \underline{r^3 + 0r^2 + 2r + 3} \\
 & \qquad - 4r^2 - 4r + 2
\end{array}
$$

$$\frac{r^4 + r^3 - 2r^2 + r + 5}{r^3 + 2r + 3} = r + 1 + \frac{-4r^2 - 4r + 2}{r^3 + 2r + 3}$$

32. $x^3 + 2x^2 + k$ has $x + 3$ as a factor if $(x^3 + 2x^2 + k) \div (x + 3)$ has a zero remainder.

$$
\begin{array}{r|l}
 & x^2 - x + 3 \\ \hline
x + 3 & x^3 + 2x^2 + 0x + k \\
 & \underline{x^3 + 3x^2} \\
 & \quad - x^2 + 0x \\
 & \quad \underline{- x^2 - 3x} \\
 & \qquad 3x + k \\
 & \qquad \underline{3x + 9} \\
 & \qquad\quad k - 9
\end{array}
$$

The remainder, $k - 9$, must equal zero. Hence, $k = 9$.

EXERCISE 3.4

In Exercises 2-24, first write each given number or expression in prime factor form.

2. $3, 2^2, 5$; 3 and 5 occur as factors at most once and 2 occurs as a factor at most twice. Hence, LCM $= 3 \cdot 5 \cdot 2^2 = 60$.

4. $2^2, 3 \cdot 5, 2 \cdot 3^2$; in any one number, 2 and 3 occur as factors at most twice and 5 occurs at most once. Hence, LCM $= 2^2 \cdot 3^2 \cdot 5 = 180$.

6. $2 \cdot 2, 11, 2 \cdot 11$; in any one number, 2 occurs as a factor at most twice and 11 occurs at most once. Hence, LCM $= 2^2 \cdot 11 = 44$.

8. $2^2 \cdot 3 \cdot x \cdot y, 2^3 \cdot 3 \cdot x^3 \cdot y^2$; in any one expression, 2 and x occur as factors at most three times, y occurs at most twice, and 3 occurs at most once. Hence, LCM $= 2^3 \cdot 3 \cdot x^3 \cdot y^2 = 24x^3y^2$.

10. $7 \cdot x, 2^3 \cdot y, 2 \cdot 3 \cdot z$; in any one expression, 7, 3, x, y, and z occur as factors at most once and 2 occurs at most three times. Hence, LCM $= 2^3 \cdot 3 \cdot 7 \cdot x \cdot y \cdot z = 168xyz$.

12. $2 \cdot 3 \cdot (x + y)^2, 2^2 \cdot x \cdot y^2$; in any one expression, 3 and x occur as factors at most once and 2, $(x + y)$, and y occur at most twice. Hence, LCM $= 2^2 \cdot 3 \cdot x \cdot y^2 \cdot (x + y)^2 = 12xy^2(x + y)^2$.

14. $x + 2, (x + 2)(x - 2)$; in any one expression, $x + 2$ and $x - 2$ occur as factors at most once. Hence, LCM $= (x + 2)(x - 2)$.

16. $(x - 1)(x - 2), (x - 1)^2$; in any one expression, $x - 1$ occurs as a factor at most twice and $x - 2$ occurs once. Hence, LCM $= (x - 1)^2(x - 2)$.

18. $(x - 2)(x + 1), (x - 2)^2$; in any one expression, $x - 2$ occurs as a factor at most twice and $x + 1$ occurs at most once. Hence. LCM $= (x - 2)^2(x + 1)$.

20. $y(y + 2), (y + 2)^2$; in any one expression, y occurs as a factor at most once and $y + 2$ occurs at most twice. Hence, LCM $= y(y + 2)^2$.

22. $3(x - 1)(x + 1)$, $(x - 1)^2$, 2^2; in any one expression, 3 and $x + 1$ occur as factors at most once and 2 and $x - 1$ occur at most twice. Hence,
LCM $= 3 \cdot 2^2 \cdot (x + 1) \cdot (x - 1)^2 = 12(x + 1)(x - 1)^2$.

24. y, $y(y + 1)(y - 1)$, $(y - 1)^3$; in any one expression, y and $y + 1$ occur as factors at most once and $y - 1$ occurs three times. Hence, LCM $= y(y + 1)(y - 1)^3$.

26. $\dfrac{y - 5}{7}$

28. $\dfrac{x - 2y + z}{3}$

30. $\dfrac{y + 1 + y - 1}{b} = \dfrac{2y}{b}$

32. $\dfrac{2 - (b - 2) + b}{a - 3b}$
$= \dfrac{2 - b + 2 + b}{a - 3b} = \dfrac{4}{a - 3b}$

34. $\dfrac{(x + 4) + (2x - 3)}{x^2 - x + 2}$
$= \dfrac{3x + 1}{x^2 - x + 2}$

36. $\dfrac{5}{6} - \dfrac{1}{5} = \dfrac{5 \cdot 5}{6 \cdot 5} - \dfrac{1 \cdot 6}{5 \cdot 6}$
$= \dfrac{25}{30} - \dfrac{6}{30} = \dfrac{19}{30}$

38. $\dfrac{2}{3} + \dfrac{5}{11} = \dfrac{2 \cdot 11}{3 \cdot 11} + \dfrac{5 \cdot 3}{11 \cdot 3}$
$= \dfrac{22}{33} + \dfrac{15}{33}$
$= \dfrac{37}{33}$

40. $\dfrac{14}{3} - z = \dfrac{14}{3} - \dfrac{z}{1}$
$= \dfrac{14}{3} - \dfrac{z \cdot 3}{1 \cdot 3}$
$= \dfrac{14}{3} - \dfrac{3z}{3} = \dfrac{14 - 3z}{3}$

42. LCD $= by$; $\dfrac{3}{by} - \dfrac{2(y)}{b(y)} = \dfrac{3 - 2y}{by}$

44. $\dfrac{x - 3}{2^2} + \dfrac{5 - x}{2 \cdot 5}$, LCD $= 2^2 \cdot 5 = 20$;
$\dfrac{(x - 3)(5)}{2^2(5)} + \dfrac{(5 - x)(2)}{2 \cdot 5(2)}$
$= \dfrac{(5x - 15) + (10 - 2x)}{20} = \dfrac{3x - 5}{20}$

46. LCD $= 6ab$;
$\dfrac{(3a + 2b)(2a)}{3b(2a)} + \dfrac{-(a + 2b)(b)}{6a(b)} = \dfrac{(6a^2 + 4ab) - (ab + 2b^2)}{6ab}$
$= \dfrac{6a^2 + 3ab - 2b^2}{6ab}$

48. LCD = $(y + 2)(y + 3)$;

$$\frac{2(y + 3)}{(y + 2)(y + 3)} + \frac{-3(y + 2)}{(y + 3)(y + 2)} = \frac{2(y + 3) - 3(y + 2)}{(y + 3)(y + 2)}$$

$$= \frac{2y + 6 - 3y - 6}{(y + 2)(y + 3)}$$

$$= \frac{-y}{(y + 2)(y + 3)}$$

50. LCD = $(2x + 1)(x - 2)$;

$$\frac{1(x - 2)}{(2x + 1)(x - 2)} + \frac{-3(2x + 1)}{(x - 2)(2x + 1)} = \frac{(x - 2) - 3(2x + 1)}{(x - 2)(2x + 1)}$$

$$= \frac{x - 2 - 6x - 3}{(2x + 1)(x - 2)}$$

$$= \frac{-5x - 5}{(2x + 1)(x - 2)}$$

52. LCD = $(a + 2)(a + 3)$;

$$\frac{(a + 1)(a + 3)}{(a + 2)(a + 3)} + \frac{-(a + 2)(a + 2)}{(a + 3)(a + 2)}$$

$$= \frac{(a^2 + 4a + 3) - (a^2 + 4a + 4)}{(a + 2)(a + 3)}$$

$$= \frac{a^2 + 4a + 3 - a^2 - 4a - 4}{(a + 2)(a + 3)} = \frac{-1}{(a + 2)(a + 3)}$$

54. LCD = $(2x - y)(x - 2y)$;

$$\frac{(x + 2y)(x - 2y)}{(2x - y)(x - 2y)} + \frac{-(2x + y)(2x - y)}{(x - 2y)(2x - y)}$$

$$= \frac{(x^2 - 4y^2) - (4x^2 - y^2)}{(2x - y)(x - 2y)} = \frac{x^2 - 4y^2 - 4x^2 + y^2}{(2x - y)(x - 2y)}$$

$$= \frac{-3x^2 - 3y^2}{(2x - y)(x - 2y)}$$

56. $\dfrac{1}{(b + 1)(b - 1)} + \dfrac{-1}{(b + 1)^2}$, LCD = $(b + 1)^2(b - 1)$;

$$\frac{1(b + 1)}{(b + 1)^2(b - 1)} + \frac{-1(b - 1)}{(b + 1)^2(b - 1)}$$

$$= \frac{b + 1 - (b - 1)}{(b + 1)^2(b - 1)} = \frac{b + 1 - b + 1}{(b + 1)^2(b - 1)} = \frac{2}{(b + 1)^2(b - 1)}$$

58. $\dfrac{8}{(a + 2b)(a - 2b)} + \dfrac{-2}{(a - 2b)(a - 3b)}$,

LCD = $(a + 2b)(a - 2b)(a - 3b)$;

$$\frac{8(a - 3b)}{(a + 2b)(a - 2b)(a - 3b)} + \frac{-2(a + 2b)}{(a + 2b)(a - 2b)(a - 3b)}$$

$$= \frac{8(a - 3b) - 2(a + 2b)}{(a + 2b)(a - 2\ )(a - 3b)}$$

$$= \frac{8a - 24b - 2a - 4b}{(a + 2b)(a - 2b)(a - 3b)} = \frac{6a - 28b}{(a + 2b)(a - 2b)(a - 3b)}$$

60. $\dfrac{3a}{(a + 5)(a - 2)} + \dfrac{-2a}{(a - 2)(a + 3)}$,

LCD = $(a - 2)(a + 3)(a + 5)$;

$$\frac{3a(a + 3)}{(a + 5)(a - 2)(a + 3)} + \frac{-2a(a + 5)}{(a - 2)(a + 3)(a + 5)}$$

$$= \frac{3a(a + 3) - 2a(a + 5)}{(a + 5)(a - 2)(a + 3)}$$

$$= \frac{3a^2 + 9a - 2a^2 - 10a}{(a - 2)(a + 3)(a + 5)} = \frac{a^2 - a}{(a - 2)(a + 3)(a + 5)}$$

62. $\dfrac{y}{1} + \dfrac{-2y}{(y + 1)(y - 1)} + \dfrac{3}{(y + 1)}$,

LCD = $(y + 1)(y - 1)$ or $(y^2 - 1)$;

$$\frac{y(y^2 - 1)}{1(y^2 - 1)} + \frac{-2y}{y^2 - 1} + \frac{3(y - 1)}{y^2 - 1} = \frac{y^3 - y - 2y + 3y - 3}{y^2 - 1}$$

$$= \frac{y^3 - 3}{y^2 - 1}$$

64. $\dfrac{4}{(a + 2b)(a - 2b)} + \dfrac{2}{(a + 2b)(a + b)} + \dfrac{4}{(a + b)(a - 2b)}$,

LCD = $(a + b)(a + 2b)(a - 2b)$;

$$\frac{4(a + b)}{(a + 2b)(a - 2b)(a + b)} + \frac{2(a - 2b)}{(a + 2b)(a + b)(a - 2b)}$$

$$+ \frac{4(a + 2b)}{(a + b)(a - 2b)(a + 2b)}$$

$$= \frac{4a + 4b + 2a - 4b + 4a + 8b}{(a + b)(a + 2b)(a - 2b)} = \frac{10a + 8b}{(a + b)(a + 2b)(a - 2b)}$$

66. LCD $= (a + b)(a + c)(b + c)$;

$$\frac{1(a + c)}{(a + b)(a + c)(b + c)} + \frac{-1(a + b)}{(a + b)(a + c)(b + c)} + \frac{-1(b + c)}{(a + b)(a + c)(b + c)}$$

$$= \frac{a + c - a - b - b - c}{(a + b)(a + c)(b + c)} = \frac{-2b}{(a + b)(a + c)(b + c)}$$

68. $y^3 - 64 = y^3 - 4^3 = (y - 4)(y^2 + 4y + 16)$

Hence, LCD $= (y - 4)(y^2 + 4y + 16)$.

$$\frac{(y - 4) \cdot 1}{(y - 4)(y^2 + 4y + 16)} + \frac{y}{(y - 4)(y^2 + 4y + 16)}$$

$$= \frac{y - 4 + y}{(y - 4)(y^2 + 4y + 16)} = \frac{2y - 4}{y^3 - 64}$$

EXERCISE 3.5

2. $\frac{4}{3 \cdot 5} \cdot \frac{3}{4 \cdot 4}$

$$= \frac{1(3 \cdot 4)}{4 \cdot 5(3 \cdot 4)} = \frac{1}{20}$$

4. $\frac{7}{8} \cdot \frac{3 \cdot 16}{4 \cdot 16}$

$$= \frac{21(16)}{32(16)} = \frac{21}{32}$$

6. $\frac{3}{10} \cdot \frac{4 \cdot 4}{3 \cdot 3 \cdot 3} \cdot \frac{3 \cdot 10}{4 \cdot 9}$

$$= \frac{4(3 \cdot 3 \cdot 4 \cdot 10)}{3 \cdot 9(3 \cdot 3 \cdot 4 \cdot 10)}$$

$$= \frac{4}{27}$$

8. $\frac{2 \cdot 2 \cdot a \cdot a}{3} \cdot \frac{3 \cdot 2 \cdot b}{2 \cdot a}$

$$= \frac{4ab(2 \cdot 3 \cdot a)}{(2 \cdot 3 \cdot a)}$$

$$= 4ab$$

10. $\frac{3 \cdot 7 \cdot t \cdot t}{5 \cdot s} \cdot \frac{3 \cdot 5 \cdot s \cdot s \cdot s}{7 \cdot s \cdot t} = \frac{3 \cdot 3 \cdot s \cdot t(5 \cdot 7 \cdot t \cdot s \cdot s)}{(5 \cdot 7 \cdot t \cdot s \cdot s)}$

$$= 9st$$

12. $\frac{2 \cdot 7 \cdot a^2 \cdot a \cdot b}{3 \cdot b} \cdot \frac{-2 \cdot 3}{7 \cdot a^2} = \frac{-2 \cdot 2 \cdot a(3 \cdot 7 \cdot a^2 \cdot b)}{(3 \cdot 7 \cdot a^2 \cdot b)} = -4a$

14. $\frac{3ax^2(axy)}{4(axy)} = \frac{3ax^2}{4}$

16. $\dfrac{2 \cdot 5x}{2 \cdot 6y} \cdot \dfrac{3x^2z}{5x^3z} \cdot \dfrac{6y^2x}{3yz}$

$= \dfrac{x(2 \cdot 3 \cdot 5 \cdot 6x^3y^2z)}{z(2 \cdot 3 \cdot 5 \cdot 6x^3y^2z)}$

$= \dfrac{x}{z}$

18. $\dfrac{15x^2y \cdot 3}{45xy^2}$

$= \dfrac{x(3 \cdot 3 \cdot 5x)}{y(3 \cdot 3 \cdot 5x)} = \dfrac{x}{y}$

20. $\dfrac{3y}{2y(2x - 3y)} \cdot \dfrac{(2x - 3y)}{12x} = \dfrac{1[3y(2x - 3y)]}{8x[3y(2x - 3y)]} = \dfrac{1}{8x}$

22. $\dfrac{(3x + 5)(3x - 5)}{2(x - 1)} \cdot \dfrac{(x + 1)(x - 1)}{2(3x - 5)}$

$= \dfrac{(3x + 5)(x + 1)[(3x - 5)(x - 1)]}{4[3x - 5)(x - 1)]} = \dfrac{(3x + 5)(x + 1)}{4}$

24. $\dfrac{(2x + 3)(2x + 1)}{(2x - 3)(x - 1)} \cdot \dfrac{3x(2x - 3)}{(1 + 2x)(1 - 2x)}$

$= \dfrac{3x(2x + 3)[(2x + 1)(2x - 3)]}{(x - 1)(1 - 2x)[(2x + 1)(2x - 3)]} = \dfrac{3x(2x + 3)}{(x - 1)(1 - 2x)}$

26. $\dfrac{(x - 2)(x + 1)}{(x + 3)(x + 1)} \cdot \dfrac{(x + 1)(x - 5)}{(x - 5)(x + 2)}$

$= \dfrac{(x - 2)(x + 1)[(x + 1)(x - 5)]}{(x + 3)(x + 2)[(x + 1)(x - 5)]} = \dfrac{(x - 2)(x + 1)}{(x + 3)(x + 2)}$

28. $\dfrac{(x - 3)(3x + 2)}{(2x + 1)(x - 1)} \cdot \dfrac{(2x + 1)(x - 5)}{(3x + 2)(x - 5)}$

$= \dfrac{(x - 3)[(3x + 2)(2x + 1)(x - 5)]}{(x - 1)[(3x + 2)(2x + 1)(x - 5)]} = \dfrac{x - 3}{x - 1}$

30. $\dfrac{5x(x - 1)}{3} \cdot \dfrac{(x + 1)(x - 10)}{4(x - 10)} \cdot \dfrac{y^2}{-2(x + 1)(x - 1)}$

$= \dfrac{5xy^2[(x + 1)(x - 1)(x - 10)]}{-24[(x + 1)(x - 1)(x - 10)]} = \dfrac{-5xy^2}{24}$

32. $\dfrac{2}{3} \cdot \dfrac{15}{9} = \dfrac{10(3)}{9(3)} = \dfrac{10}{9}$

34. $\dfrac{9ab^3}{x} \cdot \dfrac{2x^3}{3} = \dfrac{6ab^3x^2(3x)}{1(3x)}$

$= 6ab^3x^2$

36. $\dfrac{24a^3b}{1} \cdot \dfrac{7x}{3a^2b}$

$= \dfrac{56ax(3a^2b)}{1(3a^2b)} = 56ax$

38. $\dfrac{3(2y - 9)}{5x} \cdot \dfrac{1}{2(2y - 9)}$

$= \dfrac{3(2y - 9)}{10x(2y - 9)} = \dfrac{3}{10x}$

40. $\frac{(a+5)(a-3)}{(a+5)(a-2)} \cdot \frac{(a-2)(a-7)}{(a+3)(a-3)}$

$= \frac{(a-7)[(a+5)(a-3)(a-2)]}{(a+3)[(a+5)(a-3)(a-2)]} = \frac{a-7}{a+3}$

42. $\frac{(x+7)(x-1)}{(x+2)(x-1)} \cdot \frac{(x-5)(x+2)}{(x-2)(x+7)}$

$= \frac{(x-5)[(x-1)(x+2)(x+7)]}{(x-2)[\ x-1)(x+2)(x+7)]} = \frac{x-5}{x-2}$

44. $\frac{(3x+2)(3x-1)}{(4x-1)(3x+2)} \cdot \frac{(4x-1)(2x+3)}{(3x-1)(3x-1)}$

$= \frac{(2x+3)[(3x-1)(4x-1)(3x+2)]}{(3x-1)[(3x-1)(4x-1)(3x+2)]} = \frac{2x+3}{3x-1}$

46. $\frac{(2x-y)(4x^2+2xy+y^2)}{(x+y)} \cdot \frac{(x+y)(x-y)}{(2x-y)}$

$= \frac{(x-y)(4x^2+2xy+y^2)[(x+y)(2x-y)]}{1[x+y)(2x-y)]}$

$= (x-y)(4x^2+2xy+y^2)$

48. $\frac{(2x+3)(y+2)}{(2x+3)} \cdot \frac{(y-1)}{(y+2)} = \frac{(y-1)[(2x+3)(y+2]}{1[(2x+3)(y+2)]} = y-1$

50. $\frac{(y-3)(y-1)}{y^2} \cdot \frac{y(y+1)}{(y-3)^2} \cdot \frac{(y-3)(y+2)}{(y-3)(y+1)}$

$= \frac{(y-1)(y+2)[y(y+1)(y-3)^2]}{y(y-3)[y(y+1)(y-3)^2]} = \frac{(y-1)(y+2)}{y(y-3)}$

52. $Q(x) \div R(x) = \frac{x^2}{x^2-1} \div \frac{x^3}{(x-1)^2} = \frac{x^2}{(x+1)(x-1)} \cdot \frac{(x-1)^2}{x^3}$

$= \frac{(x-1)[x^2(x-1)]}{x(x+1)[x^2(x-1)]} = \frac{x-1}{x(x+1)}$

54. $P(x) \cdot R(x) \div Q(x) = \frac{x}{x-1} \cdot \frac{x^3}{(x-1)^2} \div \frac{x^2}{x^2-1}$

$= \frac{x}{x-1} \cdot \frac{x^3}{(x-1)^2} \cdot \frac{(x-1)(x+1)}{x^2}$

$= \frac{x^2(x+1)[x^2(x-1)]}{(x-1)^2[x^2(x-1)]} = \frac{x^2(x+1)}{(x-1)^2}$

EXERCISE 3.6

2. $\frac{4}{5} \cdot \frac{5}{7} = \frac{4 \cdot 5}{7 \cdot 5} = \frac{4}{7}$

4. $\frac{5}{2} \cdot \frac{4}{21} = \frac{5 \cdot 2 \cdot 2}{21 \cdot 2} = \frac{10}{21}$

6. $\frac{5x}{6y} \cdot \frac{5y}{4x} = \frac{5 \cdot 5 \cdot [x \cdot y]}{6 \cdot 4 \cdot [x \cdot y]}$

$= \frac{25}{24}$

8. $\frac{3ab}{4} \cdot \frac{8a^2}{3b}$

$= \frac{2a^3 \cdot [4 \cdot 3 \cdot b]}{1 \cdot [4 \cdot 3 \cdot b]} = 2a^3$

10. $4 + \frac{2}{3} = \frac{14}{3}$;

$\frac{\frac{1}{3}}{\frac{14}{3}} = \frac{1}{3} \cdot \frac{3}{14} = \frac{1 \cdot 3}{14 \cdot 3}$

$= \frac{1}{14}$

or

$\frac{\frac{1}{3}\left(\frac{3}{1}\right)}{\left(4 + \frac{2}{3}\right)\left(\frac{3}{1}\right)} = \frac{1}{12 + 2} = \frac{1}{14}$

12. $\frac{\left(\frac{1}{2} + \frac{3}{4}\right)\left(\frac{4}{1}\right)}{\left(\frac{1}{2} - \frac{3}{4}\right)\left(\frac{4}{1}\right)} = \frac{2 + 3}{2 - 3} = -5$

14. $\frac{\left(1 + \frac{2}{a}\right)\left(\frac{a^2}{1}\right)}{\left(1 - \frac{4}{a^2}\right)\left(\frac{a^2}{1}\right)}$

$= \frac{a^2 + 2a}{a^2 - 4}$

$= \frac{a(a + 2)}{(a - 2)(a + 2)} = \frac{a}{a - 2}$

16. $\frac{\left(1 + \frac{1}{x}\right)\left(\frac{x}{1}\right)}{\left(1 - \frac{1}{x}\right)\left(\frac{x}{1}\right)} = \frac{x + 1}{x - 1}$

18. $\frac{(4)x}{\left(\frac{2}{x} + 2\right)\left(\frac{x}{1}\right)}$

$= \frac{4x}{2 + 2x} = \frac{2 \cdot 2x}{2(1 + x)}$

$= \frac{2x}{1 + x}$

20. $\frac{(y + 3)\left(\frac{y}{1}\right)}{\left(\frac{9}{y} - y\right)\left(\frac{y}{1}\right)} = \frac{y(y + 3)}{9 - y^2}$

$= \frac{y(3 + y)}{(3 - y)(3 + y)}$

$= \frac{y}{3 - y}$

22. $1 - \frac{1 \cdot (3)}{\left(1 - \frac{1}{3}\right)\left(\frac{3}{1}\right)} = 1 - \frac{3}{3 - 1} = 1 - \frac{3}{2} = \frac{-1}{2}$

24. $x - \dfrac{x(1 - x)}{\left(1 - \dfrac{x}{1 - x}\right)\left(\dfrac{1 - x}{1}\right)} = x - \dfrac{x(1 - x)}{(1 - x) - x} = x - \dfrac{x - x^2}{1 - 2x}$

$= \dfrac{x(1 - 2x)}{1 - 2x} - \dfrac{(x - x^2)}{1 - 2x}$

$= \dfrac{x(1 - 2x) - (x - x^2)}{1 - 2x}$

$= \dfrac{-x^2}{1 - 2x}$ or $\dfrac{x^2}{2x - 1}$

26. $2y + \dfrac{3(y - 1)}{\left(3 - \dfrac{2y}{y - 1}\right)\left(\dfrac{y - 1}{1}\right)} = 2y + \dfrac{3y - 3}{3(y - 1) - 2y} = 2y + \dfrac{3y - 3}{y - 3}$

$= \dfrac{2y(y - 3)}{y - 3} + \dfrac{3y - 3}{y - 3}$

$= \dfrac{2y(y - 3) + (3y - 3)}{y - 3}$

$= \dfrac{2y^2 - 3y - 3}{y - 3}$

28. Simplifying the numerator, we have:

$1 - \dfrac{1}{\dfrac{a}{b} + 2} = 1 - \dfrac{(1)b}{\left(\dfrac{a}{b} + 2\right)\left(\dfrac{b}{1}\right)} = 1 - \dfrac{b}{a + 2b} = \dfrac{(a + 2b) - b}{a + 2b}$

$= \dfrac{a + b}{a + 2b}$

Simplifying the denominator, we have:

$1 + \dfrac{3}{\dfrac{a}{2b} + 1} = 1 + \dfrac{(3)2b}{\left(\dfrac{a}{2b} + 1\right)\left(\dfrac{2b}{1}\right)} = 1 + \dfrac{6b}{a + 2b} = \dfrac{(a + 2b) + 6b}{a + 2b}$

$= \dfrac{a + 8b}{a + 2b}$

Then

$\dfrac{\dfrac{a + b}{a + 2b}\left(\dfrac{a + 2b}{1}\right)}{\dfrac{a + 8b}{a + 2b}\left(\dfrac{a + 2b}{1}\right)} = \dfrac{a + b}{a + 8b}.$

30. $\dfrac{\left(a + 4 - \dfrac{7}{a - 2}\right)\left(\dfrac{a - 2}{1}\right)}{\left(a - 1 + \dfrac{2}{a - 2}\right)\left(\dfrac{a - 2}{1}\right)} = \dfrac{(a + 4)(a - 2) - 7}{(a - 1)(a - 2) + 2}$

$= \dfrac{a^2 + 2a - 15}{a^2 - 3a + 4}$

32. $$\frac{\left(\frac{a}{bc} - \frac{b}{ac} + \frac{c}{ab}\right)\left(\frac{a^2b^2c^2}{1}\right)}{\left(\frac{1}{a^2b^2} - \frac{1}{a^2c^2} + \frac{1}{b^2c^2}\right)\left(\frac{a^2b^2c^2}{1}\right)} = \frac{a^3bc - ab^3c + abc^3}{c^2 - b^2 + a^2}$$

$$= \frac{abc(a^2 - b^2 + c^2)}{(a^2 - b^2 + c^2)} = abc$$

34. $$\frac{\left(\frac{a + 1}{a - 1} + \frac{a - 1}{a + 1}\right)(a + 1)(a - 1)}{\left(\frac{a + 1}{a - 1} - \frac{a - 1}{a + 1}\right)(a + 1)(a - 1)}$$

$$= \frac{(a + 1)^2 + (a - 1)^2}{(a + 1)^2 - (a - 1)^2} = \frac{(a^2 + 2a + 1) + (a^2 - 2a + 1)}{(a^2 + 2a + 1) - (a^2 - 2a + 1)}$$

$$= \frac{2a^2 + 2}{4a} = \frac{2(a^2 + 1)}{2(2a)} = \frac{a^2 + 1}{2a}$$

4

FIRST-DEGREE EQUATIONS AND INEQUALITIES

EXERCISE 4.1

2. $x - 2 + \boxed{(2)} = 8 + \boxed{(2)}$

   $x = 10; \quad \{10\}$

4. $7 + \boxed{(-4)} = x + 4 + \boxed{(-4)}$

   $3 = x; \quad \{3\}$

6. $x - 6 + \boxed{(-x)} = 2x + \boxed{(-x)}$

   $-6 = x; \quad \{-6\}$

8. $4x - 3 + \boxed{(-2x) + (3)} = 2x + 1 + \boxed{(-2x) + (3)}$

   $2x = 4; \quad \{2\}$

10. $3x + 12 = 6$

    $3x + 12 + \boxed{(-12)} = 6 + \boxed{(-12)}$

    $3x = -6; \quad \{-2\}$

12. $3x - 3 - x = 0$

    $2x - 3 + \boxed{(3)} = 0 + \boxed{(3)}$

    $2x = 3; \quad \left\{\frac{3}{2}\right\}$

14. $4[2x + 3x - 1] = 5$

    $4[5x - 1] = 5$

    $20x - 4 = 5$

    $20x - 4 + \boxed{(4)} = 5 + \boxed{(4)}$

    $20x = 9; \quad \left\{\frac{9}{20}\right\}$

16. $-5[2x - 2x - 2] = 6 - x$

$$-5[-2] = 6 - x$$
$$10 = 6 - x$$
$$10 + \boxed{(-6)} = 6 - x + \boxed{(-6)}$$
$$4 = -x; \quad \{-4\}$$

18. $$2x^2 + x - 6 = x + 4 + 2x^2$$
$$2x^2 + x - 6 + \boxed{(-2x^2) + (-x)} = x + 4 + 2x^2 + \boxed{(-2x^2) + (-x)}$$
$$-6 = 4; \quad \emptyset$$

20. $$4x^2 - 12x + 9 = 4x^2 - 8$$
$$4x^2 - 12x + 9 \boxed{(-4x^2) + (-9)} = 4x^2 - 8 + \boxed{(-4x^2) + (-9)}$$
$$-12x = -17; \quad \left\{\frac{17}{12}\right\}$$

22. $$3x - (x^2 - 2x - 3) = 6x - x^2$$
$$3x - x^2 + 2x + 3 = 6x - x^2$$
$$5x - x^2 + 3 + \boxed{(x^2) + (-3) + (-6x)} = 6x - x^2 + \boxed{(x^2) + (-3) + (-6x)}$$
$$-x = -3; \quad \{3\}$$

24. $(3)\frac{2x}{3} = (3)8$

$2x = 24; \quad \{12\}$

26. $(7)\frac{2}{7}x = (7)(-8)$

$2x = -56; \quad \{-28\}$

28. $$(5)x + (5)4 = (5)\frac{2}{5}x - (5)3$$
$$5x + 20 = 2x - 15$$
$$5x + 20 + \boxed{(-2x) + (-20)} = 2x - 15 + \boxed{(-2x) + (-20)}$$
$$3x = -35; \quad \left\{\frac{-35}{3}\right\}$$

30. $$(15)4 + (15)\frac{x}{5} = (15)\frac{5}{3}$$
$$60 + 3x = 25$$
$$60 + 3x + \boxed{(-60)} = 25 + \boxed{(-60)}$$
$$3x = -35; \quad \left\{\frac{-35}{3}\right\}$$

32. $$\left(12\right)\frac{x}{4} = (12)2 - \left(12\right)\frac{x}{3}$$
$$3x = 24 - 4x$$
$$3x + \boxed{(4x)} = 24 - 4x + \boxed{(4x)}$$
$$7x = 24; \quad \left\{\frac{24}{7}\right\}$$

34. $$\left(6\right)\frac{2x}{3} - \left(6\right)\frac{(2x + 5)}{6} = \left(6\right)\frac{1}{2}$$
$$4x - (2x + 5) = 3$$
$$4x - 2x - 5 = 3$$
$$2x - 5 + \boxed{5} = 3 + \boxed{5}$$
$$2x = 8; \quad \{4\}$$

36. $$(x - 9)(x + 12)\frac{2}{(x - 9)} = (x - 9)(x + 12)\frac{9}{(x + 12)}$$
$$2(x + 12) = 9(x - 9) \qquad (x \neq -12,9)$$
$$2x + 24 = 9x - 81$$
$$2x + 24 + \boxed{[-9x - 24]} = 9x - 81 + \boxed{[-9x - 24]}$$
$$-7x = -105; \quad \{15\}$$

38. $$(x - 3)\frac{5}{(x - 3)} = (x - 3)\frac{x + 2}{x - 3} + (x - 3)3 \qquad (x \neq 3)$$
$$5 = x + 2 + 3x - 9$$
$$5 = 4x - 7$$
$$5 + \boxed{7} = 4x - 7 + \boxed{7}$$
$$12 = 4x$$
$$3 = x$$

The solution set is $\emptyset$, since 3 will not satisfy the original equation.

40. $$\frac{y}{y + 2} - \frac{3}{y - 2} = \frac{y^2 + 8}{(y + 2)(y - 2)}$$
$$(y + 2)(y - 2)\frac{y}{(y + 2)} - (y + 2)(y - 2)\frac{3}{(y - 2)}$$
$$= (y + 2)(y - 2)\frac{y^2 + 8}{(y + 2)(y - 2)} \qquad (x \neq -2,2)$$
$$y(y - 2) - 3(y + 2) = y^2 + 8$$
$$y^2 - 2y - 3y - 6 = y^2 + 8$$
$$y^2 - 5y - 6 + \boxed{(-y^2) + 6} = y^2 + 8 + \boxed{(-y^2) + 6}$$
$$-5y = 14; \quad \left\{\frac{-14}{5}\right\}$$

42. If {-1} is the solution set, then the equation must be satisfied if -1 is substituted for x.

$$2(-1) - 3 = \frac{4 + (-1)}{k}, \ -5 = \frac{3}{k}, \ -5k = 3, \ k = \frac{-3}{5}$$

44. If {-7} is the solution set, then the equation will be satisfied if -7 is substituted for x.

$$\frac{(-7) - 3}{3(-7) + k} + 1 = \frac{12}{7}$$

$$\frac{-10}{-21 + k} + 1 = \frac{12}{7}$$

$$7(-21 + k) \cdot \frac{-10}{-21 + k} + 7(-21 + k) \cdot 1 = 7(-21 + k) \cdot \frac{12}{7}$$

$$-70 + (-147) + 7k = -252 + 12k$$

$$-217 + 7k + \boxed{217 + (-12k)} = -252 + 12k + \boxed{217 + (-12k)}$$

$$-5k = -35$$

$$k = 7$$

EXERCISE 4.2

2. $12 = \dfrac{a}{1 - \frac{1}{2}}$

$12 = \dfrac{a}{\frac{1}{2}}$

$12 = 2a$

$6 = a$

4. $58 = -2 + (n - 1)12$

$58 = -2 + 12n - 12$

$58 = -14 + 12n$

$72 = 12n$

$6 = n$

6. $\frac{1}{3} = \frac{1}{s} + \frac{1}{4}$

$\frac{1}{3}(12s) = \frac{1}{s}(12s) + \frac{1}{4}(12s)$

$4s = 12 + 3s$

$s = 12$

8. $C = 3; \quad S = 5000$

$5000 = \dfrac{15A}{3}$

$5000 = 5A$

$1000 = A$

$1000 should be allocated to advertising

10. $P = 3000$; $r = 0.075$; $A = 3900$

$$3900 = 3000 + 3000(0.075)t$$
$$3900 = 3000 + 225t$$
$$900 = 225t$$
$$4 = t$$

It would take 4 years.

12. $t = 2$; $E = 120$; kwh $= 2.88$

$$2.88 = \frac{120I(2)}{1000}$$
$$2.88 = \frac{240I}{1000}$$
$$2880 = 240I$$
$$12 = I$$

I equals 12 amperes.

14. $D = 4800$

$$4800 = -60P + 12{,}000$$
$$-7200 = -60P$$
$$120 = P$$

The price should be \$120.

16. $s = 165$; hp $= 468$

$$468 = \frac{62.4N(165)}{33{,}000}$$
$$468 = \frac{10{,}296N}{33{,}000}$$
$$15{,}444{,}000 = 10{,}296N$$
$$1500 = N$$

1500 cubic feet of water flowed.

18. $n = 0.5$; $d = 0.09$; $A = 17{,}600$

$$17{,}600 = S[1 - 0.09(0.5)]$$
$$17{,}600 = S(1 - 0.045)$$
$$17{,}600 = 0.955S$$
$$18{,}429.32 = S$$

The maturity value is \$18,429.32

20. $R_1 = 20$; $R_2 = 40$; $R_n = 10$

$$\frac{1}{10} = \frac{1}{20} + \frac{1}{40} + \frac{1}{R_3}$$
$$\left(40R_3\right)\frac{1}{10} = \left(40R_3\right)\frac{1}{20} + \left(40R_3\right)\frac{1}{40} + \left(40R_3\right)\frac{1}{R_3}$$
$$4R_3 = 2R_3 + R_3 + 40$$
$$4R_3 = 3R_3 + 40$$
$$R_3 = 40$$

The third resistor must be 40 ohms.

EXERCISE 4.3

2. $\left(\frac{1}{a}\right)(b - c) = \left(\frac{1}{a}\right)(a)y$

$$\frac{b - c}{a} = y$$

4. $4by + c + \boxed{(-c)} = b + \boxed{(-c)}$

$$4by = b - c$$

$$\left(\frac{1}{4b}\right)(4b)y = \left(\frac{1}{4b}\right)(b - c)$$

$$y = \frac{b - c}{4b}$$

6. $4by + \boxed{(-2by)} = a + 2by + \boxed{(-2by)}$

$$2by = a$$

$$\left(\frac{1}{2b}\right)(2b)y = \left(\frac{1}{2b}\right)a$$

$$y = \frac{a}{2b}$$

8. $b + \boxed{(-a)} = \frac{1}{c}y + a + \boxed{(-a)}$

$$b - a = \frac{1}{c}y$$

$$(c)(b - a) = (c)\left(\frac{1}{c}\right)y$$

$$c(b - a) = y \quad \text{or}$$

$$y = cb - ca$$

10. $b(c) = \frac{(y - a)}{c}(c)$

$$bc = y - a$$

$$bc + \boxed{(a)} = y - a + \boxed{(a)}$$

$$bc + a = y$$

12. $$(a)\frac{(b - 3x)}{a} = (a)c$$

$$b - 3x = ac$$

$$b - 3x + \boxed{(-b)} = ac + \boxed{(-b)}$$

$$-3x = ac - b$$

$$\left(\frac{-1}{3}\right)(-3)x = \frac{-1}{3}(ac - b)$$

$$x = \frac{-(ac - b)}{3}$$

$$x = \frac{b - ac}{3}$$

14. $$y + \boxed{(-by)} = b + by + \boxed{(-by)}$$

$$y - by = b$$
$$(1 - b)y = b$$

$$\left(\frac{1}{1 - b}\right)(1 - b)y = \left(\frac{1}{1 - b}\right)b$$

$$y = \frac{b}{1 - b}$$

16. $$4y - 3y + 3b = 8b$$
$$y + 3b = 8b$$

$$y + 3b + \boxed{(-3b)} = 8b + \boxed{(-3b)}$$

$$y = 5b$$

18. $$(y - 2)(a + 3)\left(\frac{1}{a + 3}\right) = 2a\left(\frac{1}{a + 3}\right)$$

$$y - 2 = \frac{2a}{a + 3}$$

$$y = \frac{2a}{a + 3} + 2 \quad \text{or} \quad y = \frac{2a + 2(a + 3)}{a + 3}$$

$$= \frac{4a + 6}{a + 3}$$

20. $$(by)\frac{2}{y} - (by)\frac{3}{b} = 4(by)$$

$$2b - 3y = 4by$$
$$2b = 4by + 3y$$
$$2b = (4b + 3)y$$

$$\frac{2b}{4b + 3} = y$$

22. $$(6x)\frac{c}{3} + (6x)\frac{1}{x} = (6x)\frac{c}{6x}$$

$$2cx + 6 = c$$
$$2cx = c - 6$$

$$x = \frac{c - 6}{2c}$$

24. $(abx)\frac{1}{a} - (abx)\frac{1}{b} = (abx)\frac{1}{x}$

$$bx - ax = ab$$
$$(b - a)x = ab$$
$$x = \frac{ab}{b - a}$$

26. $\frac{E}{c^2} = m$

28. $I = p(rt)$

$$\frac{I}{rt} = p$$

30. $\frac{E}{I} = R$

32. $p - 2w = 21$

$$\frac{p - 2w}{2} = 1$$

34. $yz = k$

$$z = \frac{k}{y}$$

36. $\frac{W}{I^2} = R$

38. $S(1 - r) = a$

$$S - Sr = a$$
$$-Sr = a - S$$
$$r = \frac{a - S}{-S}$$
$$r = \frac{S - a}{S}$$

40. $S = 2r^2 + 2rh$

$$S - 2r^2 = 2rh$$
$$\frac{S - 2r^2}{2r} = h$$
$$h = \frac{S - 2r^2}{2r}$$

or $h = \frac{S}{2r} - r$

42. $A - 2\pi r^2 = (2\pi r)h$

$$\frac{A - 2\pi r^2}{2\pi r} = h$$

or

$$\frac{A}{2\pi r} - r = h$$

44. $(rst)\frac{1}{r} = (rst)\frac{1}{s} + (rst)\frac{2}{t}$

$$st = rt + 2rs$$
$$st = r(t + 2s)$$
$$\frac{st}{t + 2s} = r$$

46. From Problem 45, if $\frac{a}{b} = \frac{c}{d}$,

$$ad = bc$$
$$\left(\frac{1}{ac}\right)ad = \left(\frac{1}{ac}\right)bc$$
$$\frac{d}{c} = \frac{b}{a} \text{ or } \frac{b}{a} = \frac{d}{c}.$$

48. From Problem 45, if $\frac{a}{b} = \frac{c}{d}$,

$$ad = bc$$

$$\left(\frac{1}{ab}\right)ad = \left(\frac{1}{ab}\right)bc$$

$$\frac{d}{b} = \frac{c}{a}.$$

50. If $\frac{a}{b} = \frac{c}{d}$,

$$\frac{a}{b} - 1 = \frac{c}{d} - 1$$

$$\frac{a}{b} - \frac{b}{b} = \frac{c}{d} - \frac{d}{d}$$

$$\frac{a - b}{b} = \frac{c - d}{d}.$$

EXERCISE 4.4

2. a. The number: n

   b. $\frac{3}{5}n = n - 8$

   $3n = 5n - 40$

   $-2n = -40$

   $n = 20$

   The number is 20.

4. a. The cost: x

   The profit: $x - 15$

   Then since the cost plus the profit equals the selling price, $x + x - 15 = 85$

   b. $2x - 15 = 85$

   $2x = 100$

   $x = 50$

   $x - 15 = 35$

   The cost is \$50 and the profit is \$35.

6. a. An integer: x

   Next consecutive integer: $x + 1$

   Next consecutive integer: $x + 2$

   $x + (x + 2) = 146$

   b. $2x = 144$

   $x = 72, \quad x + 1 = 73, \quad x + 2 = 74$

   The integers are 72, 73, and 74.

8. a. An odd integer: x

   Next consecutive odd integer: $x + 2$

   Next consecutive odd integer: $x + 4$

   $2(x + x + 2) = 3(x + 4) - 1$

b. $4x + 4 = 3x + 11$

$x = 7, \quad x + 2 = 9, \quad x + 4 = 11$

The integers are 7, 9, and 11.

10. a. Number of dimes: x
Number of quarters: $x - 3$
Number of nickels: $3x$

$$\begin{bmatrix}\text{value of}\\ \text{dimes}\\ \text{in cents}\end{bmatrix} + \begin{bmatrix}\text{value of}\\ \text{quarters}\\ \text{in cents}\end{bmatrix} + \begin{bmatrix}\text{value of}\\ \text{nickels}\\ \text{in cents}\end{bmatrix} = \begin{bmatrix}\text{value of}\\ \text{collection}\\ \text{in cents}\end{bmatrix}$$

$$10x \quad + \quad 25(x - 3) \quad + \quad 5(3x) \quad = \quad 175$$

b. $10x + 25x - 75 + 15x = 175$

$50x - 75 = 175$

$50x = 250$

$x = 5, \quad x - 3 = 2, \quad 3x = 15$

There are 5 dimes, 2 quarters, and 15 nickels.

12. a. Number of chocolate covered bars: x
Number of sandwich bars: $50 - x$

$$\begin{bmatrix}\text{value of chocolate}\\ \text{covered bars in}\\ \text{cents}\end{bmatrix} + \begin{bmatrix}\text{value of sand-}\\ \text{wich bars in}\\ \text{cents}\end{bmatrix} = \begin{bmatrix}\text{value of all}\\ \text{the bars in}\\ \text{cents}\end{bmatrix}$$

$$18x \quad + \quad 15(50 - x) \quad = \quad 810$$

b. $18x + 750 - 15x = 810$

$3x + 750 = 810$

$3x = 60$

$x = 20$

$50 - x = 30$

He purchased 20 chocolate covered bars and 30 sandwich bars.

14. a. Amount invested at 5%: A
Amount invested at 7%: $2400 - A$

Then $0.05A$ = return on 5% investment
$0.07(2400 - A)$ = return on 7% investment

$0.05A = 0.07(2400 - A) + 12$

$5A = 7(2400 - A) + 1200$

b. $5A = 16,800 - 7A + 1200$
$12A = 18,000$
$A = 1500, \quad 2400 - A = 900$

\$1500 invested at 5%, \$900 at 7%

16. a. Amount invested at 3%: A
Total amount invested: $A + 6000$

b. $(0.03A) + (0.06 \times 6000) = 0.05(A + 6000)$
$3A + 36,000 = 5A + 30,000$
$-2A = -6000$
$A = 3000$

\$3000 invested at 3%

18. a. Measure of third angle: x
Measure of smallest angle: $x - 50$
Measure of remaining angle: $x - 25$

$(x) + (x - 50) + (x - 25) = 180$

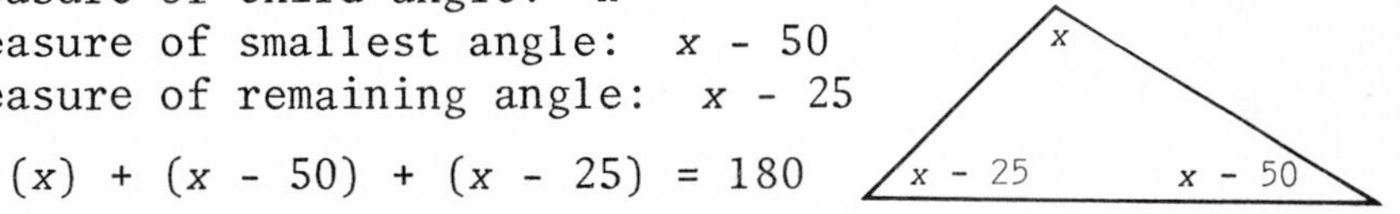

b. $3x - 75 = 180$
$3x = 255$
$x = 85, \quad x - 25 = 60, \quad x - 50 = 35$

The angles of the triangle measure 35°, 60°, and 85°

20. a. The length of a side of the original square: s
The length of a side of the larger square: $s + 5$

$(s + 5)^2 = s^2 + 85$

b. $s^2 + 10s + 25 = s^2 + 85$
$10s = 60$
$s = 6$

The length of an original side is 6 centimeters.

22. a. The distance from the fulcrum to the 24-gram weight (on the same side as the 36-gram weight): x

$36(8) + 24x = 48(8)$

b. $288 + 24x = 384$
$24x = 96, \quad x = 4$

Place the 24-gram weight 4 centimeters from the fulcrum on the same side as the 36-gram weight.

24. a. The distance from the fulcrum to the 900-lb. weight: x
The distance from the fulcrum to the 180-lb. man: $9 - x$

$900x = 180(9 - x)$

b. $$\begin{aligned} 900x &= 1620 - 180x \\ 1080x &= 1620 \\ x &= \frac{3}{2} \end{aligned}$$

The fulcrum should be $1\frac{1}{2}$ ft. (or 18 in.) from the 900-lb. weight.

26. a. Number of grams of alloy containing 45% silver: x

Alloy	Part of silver in alloy	Amount of alloy	Amount of silver
45%	0.45	x	$0.45x$
60%	0.60	$40 - x$	$0.60(40 - x)$
48%	0.48	40	$0.48(40)$

$$\begin{bmatrix} x \text{ grams} \\ 45\% \text{ alloy} \end{bmatrix} + \begin{bmatrix} (40 - x) \text{ grams} \\ 60\% \text{ alloy} \end{bmatrix} = \begin{bmatrix} 40 \text{ grams} \\ 48\% \text{ alloy} \end{bmatrix}$$

b. $$\begin{aligned} (0.45x) + 0.60(40 - x) &= 0.48(40) \\ 45x + 60(40 - x) &= 48(40) \\ 45x + 2400 - 60x &= 1920 \\ -15x &= -480 \\ x &= 32 \end{aligned}$$

Thirty-two grams of the 45% alloy.

28. a. Number of quarts of 40% antifreeze solution to be drained and replaced by water: x

Mixture	Part of antifreeze in mixture	Amount of mixture	Amount of antifreeze
40%	0.40	$32 - x$	$0.40(32 - x)$
20%	0.20	32	$0.20(32)$

$$\begin{bmatrix} \text{amount of antifreeze} \\ \text{in 40\% solution} \end{bmatrix} = \begin{bmatrix} \text{amount of antifreeze} \\ \text{in 20\% solution} \end{bmatrix}$$

$$0.40(32 - x) = 0.20(32)$$

b. $40(32 - x) = 20(32)$
$1280 - 40x = 640$
$-40x = -640$
$x = 16$

Sixteen quarts of 40% antifreeze solution should be drained and replaced with water.

30. a. Time the airplanes are in flight: t

	r	t	$d = rt$
airplane A	440	t	$440t$
airplane B	560	t	$560t$

$$\begin{bmatrix}\text{the distance} \\ \text{airplane } A \text{ flies}\end{bmatrix} + \begin{bmatrix}\text{the distance} \\ \text{airplane } B \text{ flies}\end{bmatrix} = 2500$$

$$440t + 560t = 2500$$

b. $1000t = 2500$
$t = 2.5$

The airplanes are in flight for 2.5 hours.

32. a. Distance boats have traveled when the second overtakes the first: d

	r	d	$t = d/r$
first boat	36	d	$\frac{d}{36}$
second boat	45	d	$\frac{d}{45}$

$$\begin{bmatrix}\text{time first} \\ \text{boat travels} \\ \text{in hours}\end{bmatrix} = \begin{bmatrix}\text{time second} \\ \text{boat travels} \\ \text{in hours}\end{bmatrix} + 1$$

$$\frac{d}{36} = \frac{d}{45} + 1$$

b. $180\left(\frac{d}{36}\right) = 180\left(\frac{d}{45}\right) + 180(1)$

$$5d = 4d + 180$$
$$d = 180$$

The boats will have traveled 180 nautical miles.

34. a. Time it takes for the tank to be filled: t

$$\begin{bmatrix}\text{portion of tank}\\ \text{filled by pipe 1}\end{bmatrix} - \begin{bmatrix}\text{portion of tank}\\ \text{emptied by pipe 2}\end{bmatrix} = \begin{bmatrix}\text{portion of}\\ \text{tank filled}\end{bmatrix}$$

$$\left(\frac{1}{4}\right)t \quad - \quad \left(\frac{1}{6}\right)t \quad = \quad 1$$

b. $(12)\left(\frac{1}{4}\right)t - (12)\left(\frac{1}{6}\right)t = (12)1$

$$3t - 2t = 12$$
$$t = 12$$

It will take 12 hours to fill the tank.

36. a. Time it takes second tractor to do the job alone: t

The first tractor will plow $\frac{1}{9}$ of the field in 2 hours, or $\frac{1}{18}$ of the field in 1 hour. While the two tractors work together, they plow $\frac{4}{9}$ of the field in three hours.

$$\begin{bmatrix}\text{part of field to}\\ \text{be plowed by}\\ \text{first tractor}\end{bmatrix} + \begin{bmatrix}\text{part of field to}\\ \text{be plowed by}\\ \text{second tractor}\end{bmatrix} = \begin{bmatrix}\text{part of field}\\ \text{to be plowed}\end{bmatrix}$$

$$\left(\frac{1}{18}\right)3 \quad + \quad \left(\frac{1}{t}\right)3 \quad = \quad \frac{4}{9}$$

b. $\frac{3}{18} + \frac{3}{t} = \frac{8}{18}$

$$\frac{3}{t} = \frac{5}{18}$$
$$5t = 54$$
$$t = \frac{54}{5}$$

It would take the second tractor $10\,\frac{4}{5}$ hours.

EXERCISE 4.5

2. $x + 7 + \boxed{(-7)} > 8 + \boxed{(-7)}$

$$x > 1$$

$\{x | x > 1\}$ or $(1, +\infty)$

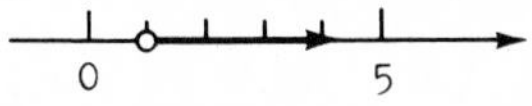

4. $2x - 3 + \boxed{(3)} < 4 + \boxed{(3)}$

$$2x < 7$$

$$\left(\frac{1}{2}\right)2x < \left(\frac{1}{2}\right)7$$

$$x < \frac{7}{2}$$

$\left\{x \middle| x < \frac{7}{2}\right\}$ or $\left(-\infty, \frac{7}{2}\right)$

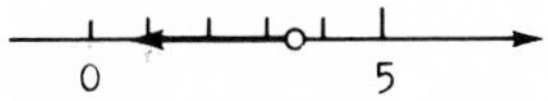

6. $2x + 3 + \boxed{(-x) + (-3)} \leqslant x - 1 + \boxed{(-x) + (-3)}$

$x \leqslant -4$; $\{x | x \leqslant -4\}$ or $(-\infty, -4]$

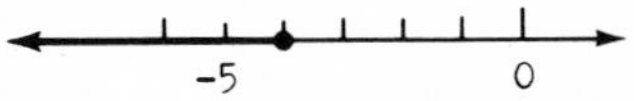

8. $(2)\frac{2x - 3}{2} \leqslant (2)5$

$$2x - 3 \leqslant 10$$

$$2x \leqslant 13$$

$$x \leqslant \frac{13}{2}$$

$\left\{x \middle| x \leqslant \frac{13}{2}\right\}$ or $\left(-\infty, \frac{13}{2}\right]$

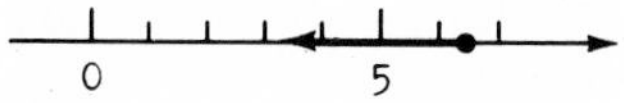

10. $\frac{-2x}{5} \leqslant 6$

$$-2x \leqslant 30$$

$$\left(-\frac{1}{2}\right)(-2)x \geqslant \left(-\frac{1}{2}\right)30$$

$x \geqslant -15$;

$\{x | x \geqslant -15\}$ or $[-15, +\infty)$

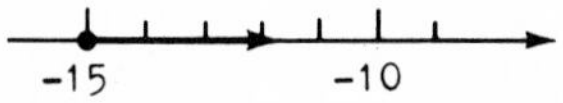

12. $\frac{-5x}{2} < -20$

$$(2)\frac{-5x}{2} < (2)(-20)$$

$$-5x < -40$$

$$\left(-\frac{1}{5}\right)(-5)x > \left(-\frac{1}{5}\right)(-40)$$

$x > 8$; $\{x | x > 8\}$ or $(8, +\infty)$

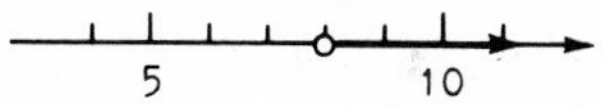

14.
$$(6)\frac{1}{2}(x + 2) \geqslant (6)\frac{2x}{3}$$
$$3(x + 2) \geqslant (2)2x$$
$$3x + 6 \geqslant 4x$$
$$3x + 6 + \boxed{(-6) + (-4x)} \geqslant 4x + \boxed{(-6) + (-4x)}$$
$$-x \geqslant -6$$
$$(-1)(-x) \leqslant (-1)(-6)$$
$$x \leqslant 6; \quad \{x | x \leqslant 6\} \quad \text{or} \quad (-\infty, 6]$$

0 5

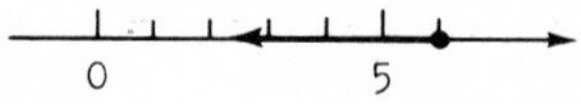

16.
$$4x - 12 \leqslant 2x + 6$$
$$4x \leqslant 2x + 18$$
$$2x \leqslant 18$$
$$x \leqslant 9$$
$$\{x | x \leqslant 9\} \quad \text{or} \quad (-\infty, 9]$$

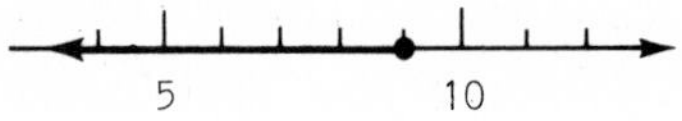

5 10

18.
$$6x + 3 \geqslant 5x - 6$$
$$6x \geqslant 5x - 9$$
$$x \geqslant -9$$
$$\{x | x \geqslant -9\} \quad \text{or} \quad [-9, +\infty)$$

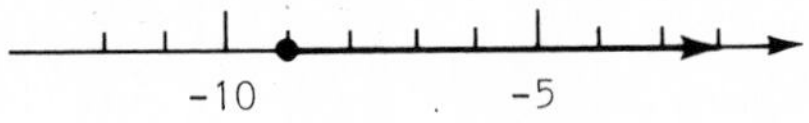

-10 -5

20.
$$0 < 2x - 6 + 5x$$
$$0 < 7x - 6$$
$$-7x < -6$$
$$x > \frac{6}{7}$$
$$\left\{x \middle| x > \frac{6}{7}\right\} \quad \text{or} \quad \left(\frac{6}{7}, +\infty\right)$$

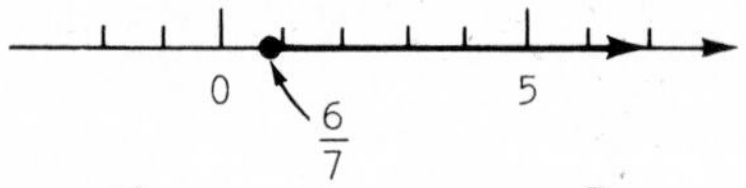

0 5 $\frac{6}{7}$

22.
$$(12)\frac{2}{3}(x - 1) + (12)\frac{3}{4}(x + 1) < (12)0$$
$$8(x - 1) + 9(x + 1) < 0$$
$$8x - 8 + 9x + 9 < 0$$
$$17x + 1 < 0$$
$$17x < -1$$
$$x < \frac{-1}{17}$$
$$\left\{x \middle| x < \frac{-1}{17}\right\} \quad \text{or} \quad \left(-\infty, -\frac{1}{17}\right)$$

-1 $\frac{-1}{17}$ 0

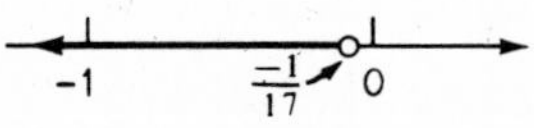

24. $(15)\frac{3}{5}(3x + 2) - (15)\frac{2}{3}(2x - 1) \leqslant (15)2$

$$9(3x + 2) - 10(2x - 1) \leqslant 30$$
$$27x + 18 - 20x + 10 \leqslant 30$$
$$7x + 28 \leqslant 30$$
$$7x \leqslant 2$$
$$x \leqslant \frac{2}{7}; \quad \left\{x \middle| x \leqslant \frac{2}{7}\right\} \text{ or } \left(-\infty, \frac{2}{7}\right]$$

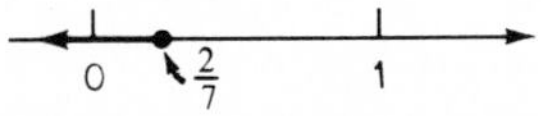

26. $\left(\frac{1}{2}\right)0 \leqslant \left(\frac{1}{2}\right)2x \leqslant \left(\frac{1}{2}\right)12$

$0 \leqslant x \leqslant 6$; $\{x | 0 \leqslant x \leqslant 6\}$ or $[0,6]$

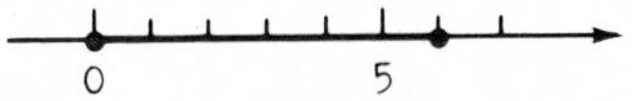

28. $2 + (4) \leqslant 3x - 4 + (4) \leqslant 8 + (4)$

$$6 \leqslant 3x \leqslant 12$$
$$\left(\frac{1}{3}\right)6 \leqslant \left(\frac{1}{3}\right)3x \leqslant \frac{1}{3}12$$

$2 \leqslant x \leqslant 4$; $\{x | 2 \leqslant x \leqslant 4\}$ or $[2,4]$

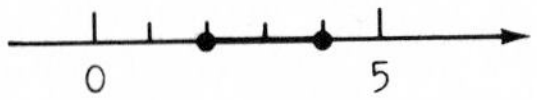

30. $-3 + (-3) < 3 - 2x + (-3) < 9 + (-3)$

$$-6 < -2x < 6$$
$$\left(-\frac{1}{2}\right)(-6) > \left(-\frac{1}{2}\right)(-2x) > \left(-\frac{1}{2}\right)6$$

$3 > x > -3$ or $-3 < x < 3$; $\{x | -3 < x < 3\}$ or $(-3,3)$

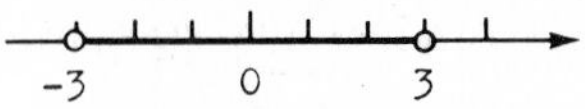

32.

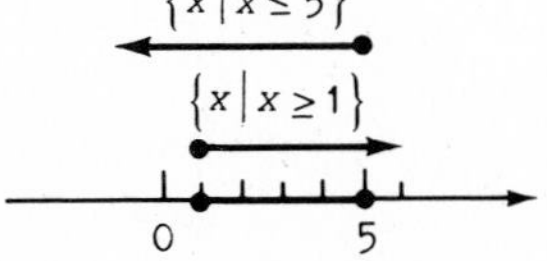

$\{x | x \leq 5\}$

$\{x | x \geq 1\}$

$\{x | x \leq 5\} \cap \{x | x \geq 1\}$

34. $2x - 3 < 5, \quad -2x + 3 < 5$

$$2x < 8 \qquad -2x < 2$$
$$x < 4 \qquad x > -1$$

The graph of the given set is the same as the graph of

$$\{x|x < 4\} \cap \{x|x > -1\}.$$

$\{x|x < 4\}$

$\{x|x > -1\}$

-1 0 5

$\{x|x < 4\} \cap \{x|x > -1\}$

36. $2 \leqslant 3x - 7 < 14, \quad -4 < 5x + 6 \leqslant 21$

$$9 \leqslant 3x < 21 \qquad -10 < 5x \leqslant 15$$
$$3 \leqslant x < 7 \qquad -2 < x \leqslant 3$$

The graph of the given set is the same as the graph of

$$\{x|3 \leqslant x < 7\} \cap \{x|-2 < x \leqslant 3\}.$$

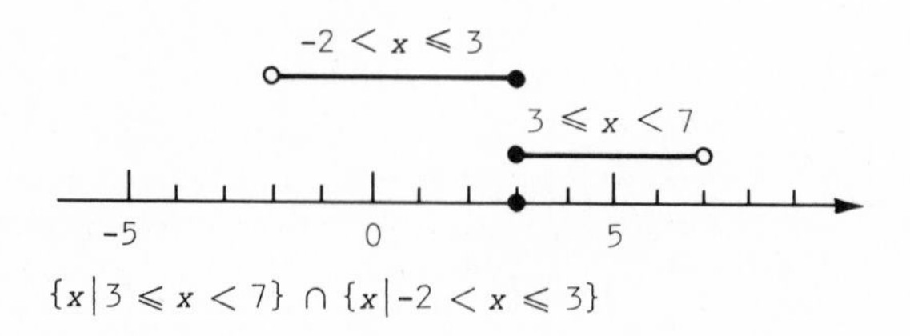

38. a. Temperature in degrees Fahrenheit: F

$$-10 < \frac{5}{9}(F - 32) < 20$$

b. $-90 < 5(F - 32) < 180$

$$-18 < F - 32 < 36$$
$$14 < F < 68$$

The temperature in Fahrenheit must be more than 14° and less than 68°.

40. a. Distance upstream: d

	d	r	$t = d/r$
upstream	d	4	$\frac{d}{4}$
downstream	d	6	$\frac{d}{6}$

$$\begin{bmatrix}\text{time sailing}\\ \text{upstream}\end{bmatrix} + \begin{bmatrix}\text{time sailing}\\ \text{downstream}\end{bmatrix} \leqslant \begin{bmatrix}\text{maximum time}\\ \text{allowed for trip}\end{bmatrix}$$

$$\frac{d}{4} + \frac{d}{6} \leqslant 10$$

b.
$$\begin{aligned} 3d + 2d &\leqslant 120 \\ 5d &\leqslant 120 \\ d &\leqslant 24 \end{aligned}$$

He can sail upstream for at most 24 miles.

EXERCISE 4.6

2. $x = 7$ or $-x = 7$

$\{7\} \cup \{-7\} = \{7, -7\}$

4. $x - 3 = 7$ or $-(x - 3) = 7$

$x = 10$ $\quad -x + 3 = 7$

$-x = 4$

$\{10\} \cup \{-4\} = \{10, -4\}$

6. $3x - 1 = 5$ or $-(3x - 1) = 5$

$3x = 6$ $\quad -3x + 1 = 5$

$x = 2$ $\quad -3x = 4$

$x = \frac{-4}{3}$

$\{2\} \cup \left\{\frac{-4}{3}\right\} = \left\{2, \frac{-4}{3}\right\}$

8. $6 - 5x = 4$ or $-(6 - 5x) = 4$

$-5x = -2$ $\quad -6 + 5x = 4$

$x = \frac{2}{5}$ $\quad 5x = 10$

$x = 2$

$\left\{\frac{2}{5}\right\} \cup \{2\} = \left\{\frac{2}{5}, 2\right\}$

10. $3x - 7 = 0$

$3x = 7$ $\quad \left\{\frac{7}{3}\right\}$

12. $x + \frac{2}{3} = \frac{1}{3}$ or $-\left(x + \frac{2}{3}\right) = \frac{1}{3}$

$x = \frac{-1}{3}$ $\qquad -x - \frac{2}{3} = \frac{1}{3}$

$-x = 1$

$x = -1$

$\left\{-\frac{1}{3}\right\} \cup \{-1\} = \left\{\frac{-1}{3}, -1\right\}$

14. $1 - \frac{1}{2}x = \frac{3}{4}$ or $-\left(1 - \frac{1}{2}x\right) = \frac{3}{4}$

$4 - 2x = 3$ $\qquad -1 + \frac{1}{2}x = \frac{3}{4}$

$-2x = -1$ $\qquad -4 + 2x = 3$

$x = \frac{1}{2}$ $\qquad 2x = 7$

$x = \frac{7}{2}$

$\left\{\frac{1}{2}\right\} \cup \left\{\frac{7}{2}\right\} = \left\{\frac{1}{2}, \frac{7}{2}\right\}$

16. $\frac{2}{3} + 2x = \frac{7}{12}$ or $-\left(\frac{2}{3} + 2x\right) = \frac{7}{12}$

$8 + 24x = 7$ $\qquad -\frac{2}{3} - 2x = \frac{7}{12}$

$24x = -1$ $\qquad -8 - 24x = 7$

$x = \frac{-1}{24}$ $\qquad -24x = 15$

$x = \frac{-15}{24} = \frac{-5}{8}$

$\left\{-\frac{1}{24}\right\} \cup \left\{-\frac{5}{8}\right\} = \left\{\frac{-1}{24}, \frac{-5}{8}\right\}$

18. $\frac{3}{4}x + \frac{3}{8} = 0$

$6x + 3 = 0$

$6x = -3$

$x = \frac{-1}{2}$ $\qquad \left\{\frac{-1}{2}\right\}$

20. $-5 < x < 5$

$\{x \mid -5 < x < 5\}$ or $(-5,5)$

-5 0 5

22. $-8 \leqslant x + 1 \leqslant 8$

$-8 + (-1) \leqslant x + 1 + (-1) \leqslant 8 + (-1)$

$-9 \leqslant x \leqslant 7$

$\{x \mid -9 \leqslant x \leqslant 7\}$ or $[-9,7]$

-10 0 10

24. $-6 < 2x + 4 < 6$

$-6 + (-4) < 2x + 4 + (-4) < 6 + (-4)$

$-10 < 2x < 2$

$\left(\frac{1}{2}\right)(-10) < \left(\frac{1}{2}\right)2x < \left(\frac{1}{2}\right)2$

$-5 < x < 1$; $\{x|-5 < x < 1\}$ or $(-5,1)$

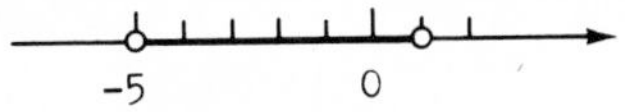

-5 0

26. $-15 \leqslant 5 - 2x \leqslant 15$

$-15 + (-5) \leqslant (-5) + 5 - 2x \leqslant 15 + (-5)$

$-20 \leqslant -2x \leqslant 10$

$\left(-\frac{1}{2}\right)(-20) \geqslant \left(-\frac{1}{2}\right)(-2x) \geqslant \left(-\frac{1}{2}\right)10$

$10 \geqslant x \geqslant -5$ or $-5 \leqslant x \leqslant 10$

$\{x|-5 \leqslant x \leqslant 10\}$ or $[-5,10]$

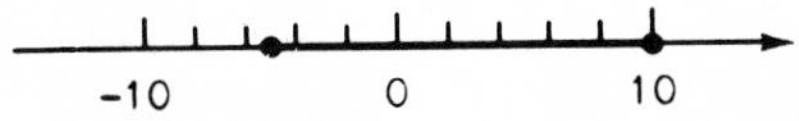

-10 0 10

28. $x \geqslant 5$ or $-x \geqslant 5$

$x \leqslant -5$; $\{x|x \leqslant -5\} \cup \{x|x \geqslant 5\}$

or $(-\infty,-5] \cup [5,+\infty)$

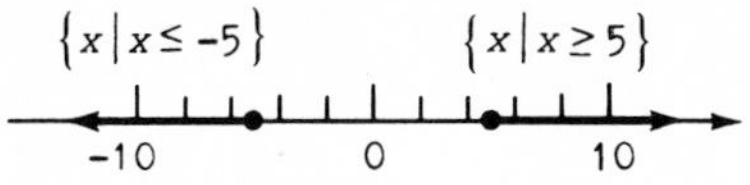

$\{x|x \leq -5\}$ $\{x|x \geq 5\}$

-10 0 10

30. $x + 5 > 2$ or $-(x + 5) > 2$

$x > -3$ $-x - 5 > 2$

$-x > 7$

$x < -7$; $\{x|x < -7\} \cup \{x|x > -3\}$

or $(-\infty,-7) \cup (-3,+\infty)$

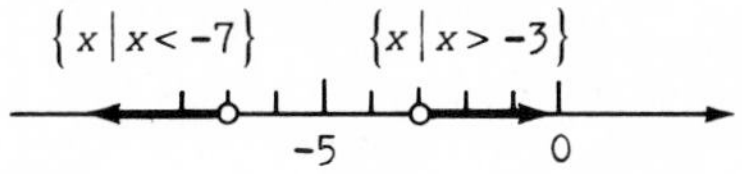

$\{x|x < -7\}$ $\{x|x > -3\}$

-5 0

32. $4 - 3x > 10$ or $-(4 - 3x) > 10$

$-3x > 6$ $-4 + 3x > 10$

$x < -2$ $3x > 14$

$x > \frac{14}{3}$; $\{x|x < -2\} \cup \left\{x|x > \frac{14}{3}\right\}$

or $(-\infty,-2) \cup \left(\frac{14}{3},+\infty\right)$

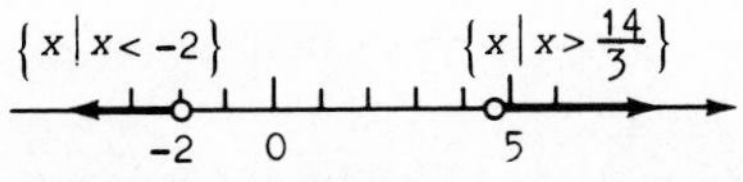

$\{x|x < -2\}$ $\left\{x|x > \frac{14}{3}\right\}$

-2 0 5

34. $|x + 2| = x + 2$ if and only if $x + 2 \geqslant 0$ or $x \geqslant -2$.

36. $|y - 5| = -(y - 5)$ if and only if $y - 5 < 0$ or $y < 5$.

38. $|3y + 4| = -(3y + 4)$ if and only if $3y + 4 \leqslant 0$ or $y \leqslant \frac{-4}{3}$.

5

EXPONENTS, ROOTS, AND RADICALS

EXERCISE 5.1

2. $y^{1+4} = y^5$

4. $b^{5+4} = b^9$

6. $\dfrac{1}{y^{6-2}} = \dfrac{1}{y^4}$

8. $x^{4-2}y^{3-1} = x^2y^2$

10. $b^{3\cdot 4} = b^{12}$

12. $y^{4\cdot 2} = y^8$

14. $x^{2\cdot 2}y^{3\cdot 2} = x^4y^6$

16. $a^{2\cdot 2}b^{3\cdot 2}c^{1\cdot 2} = a^4b^6c^2$

18. $\dfrac{y^{2\cdot 2}}{z^{3\cdot 2}} = \dfrac{y^4}{z^6}$

20. $\dfrac{3^2y^{2\cdot 2}}{x^2} = \dfrac{9y^4}{x^2}$

22. $\left[\dfrac{(-1)x^2}{2y}\right]^4 = \dfrac{(-1)^4x^{2\cdot 4}}{2^4y^4} = \dfrac{x^8}{16y^4}$

24. $(-1)^2 3^2 x^4 (-1) 5x = -1 \cdot 9 \cdot 5x^5 = -45x^5$

26. $a^6b^6(-1)^3a^3b^6 = -a^9b^{12}$

28. $\dfrac{5^2x^2}{3^3x^6} = \dfrac{25}{27x^4}$

30. $\dfrac{(-1)^2x^2y^4}{x^6y^3} = \dfrac{y}{x^4}$

32. $\dfrac{(-1)^2x^2(-1)^4x^8}{x^6} = \dfrac{x^{10}}{x^6} = x^4$

34. $\dfrac{x^4z^2(-1)^3 2^3}{2^2 \cdot x^6z^3} = \dfrac{-2}{x^2z}$

36. $\dfrac{y^2 \cdot (-1)^3 3^3}{x^2 \cdot 4^3x^3y^3} = \dfrac{-27}{64x^5y}$

38. $\left[\dfrac{a^{12}b^4c^4 \cdot x^4y^2z^2}{x^8y^4 \cdot a^2b^4c^6}\right]^2 = \left[\dfrac{a^{10}z^2}{c^2x^4y^2}\right]^2 = \dfrac{a^{20}z^4}{c^4x^8y^4}$

40. $\dfrac{m^6n^4p^2 \cdot r^3s^3 \cdot (-1)^2m^2n^2p^2}{r^4s^2 \cdot m^3n^6p^6 \cdot r^2s^2} = \dfrac{m^8n^6p^4r^3s^3}{m^3n^6p^6r^6s^4} = \dfrac{m^5}{p^2r^3s}$

42. $\dfrac{2^2x^4y^2 \cdot 2^3z^6 \cdot (-1)^2 4^2x^2z^2}{3^2 \cdot z^2 \cdot 3^3x^3y^6 \cdot 3^2y^2}$

$= \dfrac{2^5 4^2x^6y^2z^8}{3^7x^3y^8z^2} = \dfrac{512x^3z^6}{2187y^6}$

44. $\dfrac{(t - 3)^3 \cdot t}{2^3t^3 \cdot (t - 3)^2} = \dfrac{t - 3}{8t^2}$

46. $x^{2n+n-(n+1)} = x^{3n-n-1}$

$= x^{2n-1}$

48. $\left(\dfrac{y^5}{y}\right)^{2n} = (y^4)^{2n} = y^{8n}$

50. $\dfrac{y^{n(n+1)}}{y^n} = y^{(n^2+n)-n} = y^{n^2}$

EXERCISE 5.2

2. $\dfrac{1}{3^2} = \dfrac{1}{9}$

4. $3 \cdot \dfrac{1}{4^{-2}} = 3 \cdot 4^2$

$= 3 \cdot 16 = 48$

6. $(-3)^2 = 9$

8. $\dfrac{1}{3^{-2}} = 3^2 = 9$

10. $\dfrac{1^{-2}}{3^{-2}} = \dfrac{1}{3^{-2}} = 3^2 = 9$

12. $3^{-3} \cdot \dfrac{1}{6^{-2}} = \dfrac{1}{3^3} \cdot 6^2 = \dfrac{36}{27} = \dfrac{4}{3}$

14. $\dfrac{1}{5} + 1 = \dfrac{6}{5}$

16. $\dfrac{1}{8^2} - 1 = \dfrac{1}{64} - 1 = \dfrac{-63}{64}$

18. $x^3 \cdot \dfrac{1}{y^{-2}} = x^3y^2$

20. $\dfrac{1}{(xy^3)^2} = \dfrac{1}{x^2y^6}$

22. $\dfrac{2^3x^3}{y^6} = \dfrac{8x^3}{y^6}$

24. $\dfrac{3^2x^2}{y^4} \cdot \dfrac{2^2y^6}{x^2} = \dfrac{36x^2y^6}{y^4x^2}$

$= 36y^2$

26. $x^{3-(-2)}$
$= x^{3+2} = x^5$

28. 1

30. $x^{-3} \cdot \dfrac{1}{y^{-2}} = \dfrac{1}{x^3} \cdot y^2$
$= \dfrac{y^2}{x^3}$

32. $\left(\dfrac{x^{-1}y^3}{2y^{-5}}\right)^{-2} = \left(\dfrac{x^{-1}y^8}{2}\right)^{-2}$
$= \dfrac{x^2y^{-16}}{2^{-2}} = x^2y^{-16} \cdot \dfrac{1}{2^{-2}}$
$= x^2 \cdot \dfrac{1}{y^{16}} \cdot 2^2 = \dfrac{4x^2}{y^{16}}$

34. $\left(\dfrac{x^{-1}y}{xy^{-1}z}\right)^{-1} = \left(\dfrac{x^{-2}y^2}{z}\right)^{-1}$
$= \dfrac{x^2y^{-2}}{z^{-1}} = x^2y^{-2} \cdot \dfrac{1}{z^{-1}}$
$= x^2 \cdot \dfrac{1}{y^2} \cdot z = \dfrac{x^2z}{y^2}$

36. $\dfrac{2^{-1}y}{x^{-2}} \cdot \dfrac{y^2}{x} = \dfrac{2^{-1}y^3}{x^{-1}}$
$= 2^{-1}y^3 \cdot \dfrac{1}{x^{-1}}$
$= \dfrac{1}{2}y^3 \cdot x = \dfrac{xy^3}{2}$

38. $\dfrac{2^2y^4x^2}{3^2z^2} \cdot \dfrac{2^{-2}x^{-4}}{9^{-2}z^{-2}} = \dfrac{2^0x^{-2}y^4}{9 \cdot 9^{-2}z^0}$
$= \dfrac{x^{-2}y^4}{9^{-1}} = x^{-2}y^4 \cdot \dfrac{1}{9^{-1}}$
$= \dfrac{1}{x^2}y^4 \cdot 9 = \dfrac{9y^4}{x^2}$

40. $x^{-1} \cdot \dfrac{1}{y^{-1}} + \dfrac{y}{x}$
$= \dfrac{1}{x} \cdot y + \dfrac{y}{x} = \dfrac{y}{x} + \dfrac{y}{x} = \dfrac{2y}{x}$

42. $\dfrac{1}{x + y}$

44. $x \cdot \dfrac{1}{y} + \dfrac{1}{x} \cdot y = \dfrac{x}{y} + \dfrac{y}{x}$
$= \dfrac{x^2 + y^2}{xy}$

46. $\dfrac{\left(\dfrac{1}{x} + \dfrac{1}{y}\right)}{\left(\dfrac{1}{xy}\right)} = \left(\dfrac{1}{x} + \dfrac{1}{y}\right) \cdot \left(\dfrac{xy}{1}\right)$
$= y + x$

48. $x^{-n+n+1} = x$

50. $x^{n-(2n-1)}y^{n+1-n} = x^{-n+1}y$

52. $[x^{n-1-(-2)}y^{n-(-n)}]^2$
$= [x^{n+1}y^{2n}]^2 = x^{2n+2}y^{4n}$

54. $[a^{2n-(n-1)}b^{n-1-1}]^2$
$= [a^{n+1}b^{n-2}]^2 = a^{2n+2}b^{2n-4}$

56. $\dfrac{a^mb^{-n}}{d^qc^{-p}} = \dfrac{a^m}{d^q} \cdot b^{-n} \cdot \dfrac{1}{c^{-p}} = \dfrac{a^m}{d^q} \cdot \dfrac{1}{b^n} \cdot c^p = \dfrac{a^mc^p}{b^nd^q}$

58. Let $x = 1$ and $y = 2$. Then

$$(x + y)^{-2} = (1 + 2)^{-2} = \frac{1}{3^2} = \frac{1}{9}$$

and

$$\frac{1}{x^2} + \frac{1}{y^2} = \frac{1}{1^2} + \frac{1}{2^2} = 1 + \frac{1}{4} = \frac{5}{4}.$$

Therefore, $(x + y)^{-2}$ is not equivalent to $\frac{1}{x^2} + \frac{1}{y^2}$.

EXERCISE 5.3

2. 3.476×10^3
4. 6.8742×10^4
6. 4.81×10^5
8. 6.3×10^{-3}
10. 5.23×10^{-4}
12. 6×10^{-4}
14. 4,800
16. 83,100
18. 0.80
20. 0.00431
22. 143,800
24. 6.210

26. $\frac{1}{4} \times \frac{1}{10^4} = 0.25 \times 10^{-4}$
$= 0.000025$

28. $\frac{1}{5} \times \frac{1}{10^{-3}} = 0.2 \times 10^3$
$= 200$

30. $\frac{5}{8} \times \frac{1}{10^2} = 0.625 \times 10^{-2}$
$= 0.00625$

32. $\frac{10^{3-7+2}}{10^{-2+4}} = \frac{10^{-2}}{10^2} = 10^{-4}$
or 0.0001

34. $\frac{(4 \times 6) \times (10^3 \times 10^{-2})}{3 \times 10^{-7}}$
$= \frac{24 \times 10}{3 \times 10^{-7}}$
$= 8 \times 10^{1-(-7)} = 8 \times 10^8$
or 800,000,000

36. $\frac{(3^3 \times 10^3) \times (2 \times 10^{-1})}{2 \times 10^{-2}}$
$= \frac{27 \times \overset{1}{\cancel{2}} \times 10^2}{\underset{1}{\cancel{2}} \times 10^{-2}}$
$= 27 \times 10^{2-(-2)}$
$= 27 \times 10^4$ or 270,000

38. $\frac{(8^2 \times 10^8) \times (3^3 \times 10^3)}{6^2 \times 10^{-4}} = \frac{\overset{16}{\cancel{64}} \times \overset{3}{\cancel{27}} \times 10^{11}}{\underset{1}{\cancel{4}} \times \underset{1}{\cancel{9}} \times 10^{-4}}$
$= 48 \times 10^{11-(-4)} = 48 \times 10^{15}$

40. $$\frac{(65 \times 10^{-3}) \times (22 \times 10^{-1}) \times (5 \times 10)}{(13 \times 10^{-1}) \times (11 \times 10^{-3}) \times (5 \times 10^{-2})} = \frac{\overset{5}{\cancel{65}} \times \overset{2}{\cancel{22}} \times \overset{1}{\cancel{5}} \times 10^{-3}}{\underset{1}{\cancel{13}} \times \underset{1}{\cancel{11}} \times \underset{1}{\cancel{5}} \times 10^{-6}}$$

$$= 10 \times 10^{-3-(-6)} = 10 \times 10^{3} = 10^{4} \text{ or } 10{,}000$$

42. $$\frac{(54 \times 10^{-4}) \times (5 \times 10^{-2}) \times (3 \times 10^{2})}{(15 \times 10^{-4}) \times (27 \times 10^{-2}) \times (8 \times 10)} = \frac{\overset{2}{\cancel{54}} \times \overset{1}{\cancel{15}} \times 10^{-4}}{\underset{1}{\cancel{27}} \times \underset{1}{\cancel{15}} \times 8 \times 10^{-5}}$$

$$= \frac{2}{8} \times 10^{-4-(-5)}$$

$$= 0.25 \times 10 = 2.5$$

44. $$\frac{27 \times 10^{-4} \times 4 \times 10^{-3} \times 65 \times 10}{26 \times 10 \times 10^{-4} \times 9 \times 10^{-3}} = \frac{\overset{3}{\cancel{27}} \times \overset{5}{\cancel{65}} \times 4 \times \overset{1}{\cancel{10^{-6}}}}{\underset{1}{\cancel{9}} \times \underset{2}{\cancel{26}} \times \underset{1}{\cancel{10^{-6}}}}$$

$$= \frac{60}{2} = 30$$

46. In 1 hour there are 3600 seconds; in a day there are 24 hours; and in a year there are 365 days. Therefore, the number of miles in a light-year is given by

$$186{,}000 \times 3600 \times 24 \times 365$$
$$= 1.86 \times 10^{5} \times 3.6 \times 10^{3} \times 2.4 \times 10 \times 3.65 \times 10^{2}$$
$$= (1.86 \times 3.6 \times 2.4 \times 3.65) \times 10^{11}$$
$$= 58.65696 \times 10^{11}$$

EXERCISE 5.4

2. 5

4. 3

6. -3

8. $(32^{1/5})^{3} = (2)^{3} = 8$

10. $(125^{1/3})^{2} = (5)^{2} = 25$

12. $[(-64)^{1/3}]^{2} = [-4]^{2} = 16$

14. $\frac{1}{8^{1/3}} = \frac{1}{2}$

16. $(27^{1/3})^{-2} = (3)^{-2} = \frac{1}{3^2} = \frac{1}{9}$

18. $y^{(1/2)+(3/2)} = y^{4/2} = y^{2}$

20. $x^{(3/4)-(1/4)} = x^{2/4} = x^{1/2}$

22. $b^{6\cdot(2/3)} = b^{12/3} = b^{4}$

24. $y^{(-2/3)+(5/3)} = y^{3/3} = y$

26. $a^{(1/2)\cdot 6}b^{(1/3)\cdot 6}$

$= a^{6/2}b^{6/3} = a^3b^2$

28. $\dfrac{a^{(1/2)\cdot 2}}{a^{2\cdot 2}} = \dfrac{a}{a^4} = \dfrac{1}{a^3}$

30. $x^{(1/4)\cdot 8}y^{(1/2)\cdot 8} = x^2y^4$

32. $\dfrac{a^{(-1/2)\cdot 6}}{b^{(1/3)\cdot 6}} = \dfrac{a^{-3}}{b^2} = a^{-3} \cdot \dfrac{1}{b^2} = \dfrac{1}{a^3} \cdot \dfrac{1}{b^2} = \dfrac{1}{a^3b^2}$

34. $\dfrac{x^{1/2}y^{3/2}z^{-2}}{x^{-3/2}y^{1/2}} = x^{(1/2)-(-3/2)}y^{(3/2)-(1/2)} \cdot \dfrac{1}{z^2}$

$= \dfrac{x^{1/2+3/2}y^{3/2-1/2}}{z^2} = \dfrac{x^2y}{z^2}$

36. $x^{(1/5)+(2/5)} + x^{(1/5)+(4/5)} = x^{3/5} + x$

38. $x^{(3/8)+(1/4)} - x^{(3/8)+(1/2)} = x^{(3/8)+(2/8)} - x^{(3/8)+(4/8)}$

$= x^{5/8} - x^{7/8}$

40. $y^{(-1/4)+(3/4)} + y^{(-1/4)+(5/4)} = y^{1/2} + y$

42. $a^{-2/7+9/7} + a^{-2/7+2/7}$

$= a^{7/7} + a^0$

$= a + 1$

44. $x^{5/6+(-5)/6} + x^{5/6+1/6}$

$= x^0 + x^{6/6}$

$= 1 + x$

46. $\dfrac{x^{7/8}}{x^{3/8}} = x^{1/2}$; therefore

$x^{7/8} = x^{3/8}(x^{1/2})$.

48. $\dfrac{y^{-1/4}}{y} = y^{-5/4}$; therefore,

$y^{-1/4} = y(y^{-5/4})$.

50. $\dfrac{y^{3/5}}{y} = y^{(3/5)-(5/5)}$

$= y^{-2/5}$; therefore

$y^{3/5} = y(y^{-2/5})$.

52. $\dfrac{y^{1/2} + y}{y} = \dfrac{y^{1/2}}{y} + \dfrac{y}{y}$

$= y^{-1/2} + 1$; therefore,

$y^{1/2} + y = y(y^{-1/2} + 1)$.

54. $\dfrac{a^{2/3} + a^{1/3}}{a} = \dfrac{a^{2/3}}{a} + \dfrac{a^{1/3}}{a} = a^{-1/3} + a^{-2/3}$; therefore,

$a^{2/3} + a^{1/3} = a(a^{-1/3} + a^{-2/3})$

56. $a^{2\cdot(n/2)}b^{2n(2/n)} = a^nb^4$

58. $\dfrac{a^{n/2}}{b^{1/2}} \cdot \dfrac{b^{3/2}}{a^{3n}} = \dfrac{b^{(3/2)-(1/2)}}{a^{3n-(n/2)}} = \dfrac{b}{a^{5n/2}}$

60. $\left(\dfrac{m^{a}}{n^{2a}}\right)^{1/a} = \dfrac{m^{a \cdot 1/a}}{n^{2a \cdot 1/a}} = \dfrac{m}{n^2}$

62. $(x^{n+1-1}y^{n+2-2})^{1/n} = (x^n y^n)^{1/n} = x^{n \cdot 1/n} y^{n \cdot 1/n} = xy$

64. Let $a = 9$ and $b = 16$. Then

$$(a + b)^{1/2} = (9 + 16)^{1/2} = 25^{1/2} = 5$$

and

$$a^{1/2} + b^{1/2} = 9^{1/2} + 16^{1/2} = 3 + 4 = 7.$$

Therefore, $(a + b)^{1/2}$ is not equivalent to $a^{1/2} + b^{1/2}$.

EXERCISE 5.5

2. $\sqrt[3]{2}$

4. $\sqrt[3]{y^2}$

6. $5\sqrt[3]{x}$

8. $x^2\sqrt{y}$

10. $\sqrt[3]{x^2y}$

12. $-5\sqrt[3]{y^2}$

14. $\sqrt[3]{x - y}$

16. $\sqrt[4]{(3x + y)^3}$

18. $\dfrac{1}{6^{2/5}} = \dfrac{1}{\sqrt[5]{6^2}} = \dfrac{1}{\sqrt[5]{36}}$

20. $\dfrac{1}{y^{2/7}} = \dfrac{1}{\sqrt[7]{y^2}}$

22. $2^{1/3}$

24. $y^{3/2}$

26. $(ab^2)^{1/3}$ or $a^{1/3}b^{2/3}$

28. $yx^{2/3}$

30. $(a^3b)^{1/4}$ or $a^{3/4}b^{1/4}$

32. $(a + 2b)^{1/4}$

34. $b^{1/3} + 2a^{1/2}$

36. $a^{1/2}(ab)^{1/3} = a^{1/2}a^{1/3}b^{1/3} = a^{(3/6)+(2/6)}b^{1/3} = a^{5/6}b^{1/3}$

38. $\dfrac{2}{y^{1/3}}$

40. $\dfrac{5}{(a - b)^{1/3}}$

42. 12

44. -13

46. 5

48. -2

50. y

52. a^3

54. $3y^3$

56. $-a^4b^5$ 58. $\frac{3}{4}ab^2$ 60. $\frac{2}{3}ab^2$

62. $-2xy^2$ 64. $-3a^2b^3$

66.
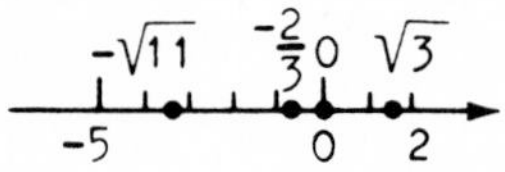

68.
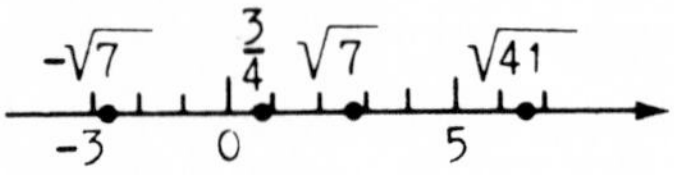

70. $3|x|y^2$ (Note that $|y^2| = y^2$.)

72. $\sqrt{(2x-1)^2} = |2x-1|$

74. $\sqrt{(x^2+y^2)^2} = x^2 + y^2$ (Note that $|x^2 + y^2| = x^2 + y^2$.)

EXERCISE 5.6

2. $\sqrt{25}\sqrt{2} = 5\sqrt{2}$

4. $\sqrt{36}\sqrt{2} = 6\sqrt{2}$

6. $\sqrt{16}\sqrt{3} = 4\sqrt{3}$

8. y^3

10. $\sqrt{y^{10}}\sqrt{y} = y^5\sqrt{y}$

12. $\sqrt{4y^4}\sqrt{y} = 2y^2\sqrt{y}$

14. $\sqrt{9y^8}\sqrt{2} = 3y^4\sqrt{2}$

16. $\sqrt[3]{y^3}\sqrt[3]{y} = y\sqrt[3]{y}$

18. $\sqrt[5]{a^{10}b^{15}}\,\sqrt[5]{a^2} = a^2b^3\sqrt[5]{a^2}$

20. $\sqrt[6]{m^6n^6}\,\sqrt[6]{m^2n} = mn\sqrt[6]{m^2n}$

22. $\sqrt[7]{4^7x^7y^7z^{14}}\,\sqrt[7]{4x^2y^3}$
$= 4xyz^2\sqrt[7]{4x^2y^3}$

24. $(\sqrt{3}\sqrt{3})\sqrt{9} = (3)3 = 9$

26. $(\sqrt{a}\sqrt{a})\sqrt{b^2} = ab$

28. $\sqrt[4]{3 \cdot 27} = \sqrt[4]{81} = 3$

30. $\sqrt{10^2}\sqrt{5 \cdot 10} = 10\sqrt{50}$
$= 10\sqrt{25}\sqrt{2}$
$= 10 \cdot 5\sqrt{2}$
$= 50\sqrt{2}$

32. $\sqrt{80 \cdot 10^4} = \sqrt{10^4}\sqrt{80}$
$= 10^2\sqrt{16}\sqrt{5}$
$= 100 \cdot 4\sqrt{5}$
$= 400\sqrt{5}$

34. $\frac{\sqrt{2}}{\sqrt{3}} = \frac{\sqrt{2}\sqrt{3}}{\sqrt{3}\sqrt{3}} = \frac{\sqrt{6}}{3}$

36. $\frac{-\sqrt{3}\sqrt{7}}{\sqrt{7}\sqrt{7}} = \frac{-\sqrt{21}}{7}$

38. $\frac{-\sqrt{y}\sqrt{3}}{\sqrt{3}\sqrt{3}} = \frac{-\sqrt{3y}}{3}$

40. $\frac{\sqrt{2a}\sqrt{b}}{\sqrt{b}\sqrt{b}} = \frac{\sqrt{2ab}}{b}$

42. $\frac{-x\sqrt{2y}}{\sqrt{2y}\sqrt{2y}} = \frac{-x\sqrt{2y}}{2y}$

44. $\frac{1\sqrt[3]{6^2x}}{\sqrt[3]{6x^2}\sqrt[3]{6^2x}} = \frac{\sqrt[3]{36x}}{\sqrt[3]{6^3x^3}} = \frac{\sqrt[3]{36x}}{6x}$

46. $\sqrt{\dfrac{x^2y^3}{y}} = \sqrt{x^2y^2} = xy$

48. $\sqrt{\dfrac{9x^3}{y^4}} = \sqrt{\dfrac{9x^2}{y^4}}\sqrt{x} = \dfrac{3x\sqrt{x}}{y^2}$

50. $\sqrt[3]{\dfrac{4r^4}{t^3}} = \sqrt[3]{\dfrac{r^3}{t^3}}\sqrt[3]{4r} = \dfrac{r\sqrt[3]{4r}}{t}$

52. $\sqrt[5]{\dfrac{x^2y^3}{xy^2}} = \sqrt[5]{xy}$

54. $\dfrac{\sqrt{2}\sqrt{2}}{3\sqrt{2}} = \dfrac{2}{3\sqrt{2}}$

56. $\dfrac{\sqrt{xy}\sqrt{xy}}{x\sqrt{xy}} = \dfrac{xy}{x\sqrt{xy}} = \dfrac{y}{\sqrt{xy}}$

58. $\sqrt[2\cdot 3]{2^{2\cdot 1}} = \sqrt[3]{2}$

60. $\sqrt[2\cdot 4]{5^{2\cdot 1}} = \sqrt[4]{5}$

62. $\sqrt[5\cdot 2]{2^{5\cdot 1}} = \sqrt{2}$

64. $\sqrt[3\cdot 3]{y^{3\cdot 1}} = \sqrt[3]{y}$

66. Least common index is 12.
$\sqrt[3]{2} = \sqrt[12]{2^4} = \sqrt[12]{16}$;
$\sqrt[4]{2} = \sqrt[12]{2^3} = \sqrt[12]{8}$;
$\sqrt[3]{2}\sqrt[4]{2} = \sqrt[12]{16}\sqrt[12]{8} = \sqrt[12]{128}$

68. Least common index is 6.
$\sqrt[3]{x} = \sqrt[6]{x^2}$; $\sqrt{x} = \sqrt[6]{x^3}$;
$\sqrt[3]{x}\sqrt{x} = \sqrt[6]{x^2}\sqrt[6]{x^3} = \sqrt[6]{x^5}$

70. Let $a = 9$ and $b = 16$. Then

$$\sqrt{a + b} = \sqrt{9 + 16} = 5 \quad \text{and} \quad \sqrt{a} + \sqrt{b} = \sqrt{9} + \sqrt{16} = 7$$

Therefore, $\sqrt{a + b}$ is not equivalent to $\sqrt{a} + \sqrt{b}$.

EXERCISE 5.7

2. $2\sqrt{2}$

4. $\sqrt{25}\sqrt{3} + 2\sqrt{3} = 5\sqrt{3} + 2\sqrt{3}$
$= 7\sqrt{3}$

6. $\sqrt{4}\sqrt{2y} - \sqrt{9}\sqrt{2y} = 2\sqrt{2y} - 3\sqrt{2y}$
$= -\sqrt{2y}$

8. $2\sqrt{4y^2}\sqrt{2z} + 3\sqrt{16y^2}\sqrt{2z} = 2(2y)\sqrt{2z} + 3(4y)\sqrt{2z}$
$= 4y\sqrt{2z} + 12y\sqrt{2z} = 16y\sqrt{2z}$

10. $\sqrt{3b} - 2\sqrt{4}\sqrt{3b} + 3\sqrt{16}\sqrt{3b} = \sqrt{3b} - 2(2)\sqrt{3b} + 3(4)\sqrt{3b}$
$= \sqrt{3b} - 4\sqrt{3b} + 12\sqrt{3b} = 9\sqrt{3b}$

12. $\sqrt[3]{27}\sqrt[3]{2} + 2\sqrt[3]{64}\sqrt[3]{2} = 3\sqrt[3]{2} + 2(4)\sqrt[3]{2} = 3\sqrt[3]{2} + 8\sqrt[3]{2} = 11\sqrt[3]{2}$

14. $10 - 5\sqrt{7}$

16. $\sqrt{36} - \sqrt{45} = 6 - \sqrt{9}\sqrt{5}$
$= 6 - 3\sqrt{5}$

18. $2 + \sqrt{2} - 2\sqrt{2} - 2 = -\sqrt{2}$

20. $(2)^2 - (\sqrt{x})^2 = 4 - x$

22. $2(3) + \sqrt{15} - 2\sqrt{15} - 5$
$= 1 - \sqrt{15}$

24. $(\sqrt{2} - 2\sqrt{3})(\sqrt{2} - 2\sqrt{3})$
$= 2 - 2\sqrt{6} - 2\sqrt{6} + 4(3)$
$= 14 - 4\sqrt{6}$

26. $5 + 5 \cdot 2\sqrt{2} = 5(1 + 2\sqrt{2})$

28. $5\sqrt{5} - 5 = 5(\sqrt{5} - 1)$

30. $3 + \sqrt{9}\sqrt{2x} = 3 + 3\sqrt{2x}$
$= 3(1 + \sqrt{2x})$

32. $\sqrt{4}\sqrt{3} - (2\sqrt{3})\sqrt{2}$
$= (2\sqrt{3}) - (2\sqrt{3})\sqrt{2}$
$= 2\sqrt{3}(1 - \sqrt{2})$

34. $\dfrac{2 \cdot 3 + 2\sqrt{5}}{2} = \dfrac{2(3 + \sqrt{5})}{2}$
$= 3 + \sqrt{5}$

36. $\dfrac{8 - 2\sqrt{4}\sqrt{3}}{4} = \dfrac{8 - 4\sqrt{3}}{4}$
$= \dfrac{4(2 - \sqrt{3})}{4} = 2 - \sqrt{3}$

38. $\dfrac{xy - x\sqrt{y^2}\sqrt{x}}{xy} = \dfrac{xy - xy\sqrt{x}}{xy}$
$= \dfrac{xy(1 - \sqrt{x}}{xy}$
$= 1 - \sqrt{x}$

40. $\dfrac{\sqrt{x} - y\sqrt{x^2}\sqrt{x}}{\sqrt{x}} = \dfrac{\sqrt{x} - xy\sqrt{x}}{\sqrt{x}}$
$= \dfrac{(1 - xy)\sqrt{x}}{\sqrt{x}}$
$= 1 - xy$

42. $\dfrac{1(2 + \sqrt{2})}{(2 - \sqrt{2})(2 + \sqrt{2})}$
$= \dfrac{2 + \sqrt{2}}{4 - 2} = \dfrac{2 + \sqrt{2}}{2}$

44. $\dfrac{2(4 + \sqrt{5})}{(4 - \sqrt{5})(4 + \sqrt{5})} = \dfrac{2(4 + \sqrt{5})}{16 - 5}$
$= \dfrac{2(4 + \sqrt{5})}{11}$
or $\dfrac{8 + 2\sqrt{5}}{11}$

46. $\dfrac{y(\sqrt{3} + y)}{(\sqrt{3} - y)(\sqrt{3} + y)}$
$= \dfrac{y(\sqrt{3} + y)}{3 - y^2}$ or
$\dfrac{y\sqrt{3} + y^2}{3 - y^2}$

48. $\dfrac{(\sqrt{x} + \sqrt{y})(\sqrt{x} + \sqrt{y})}{(\sqrt{x} - \sqrt{y})(\sqrt{x} + \sqrt{y})}$
$= \dfrac{x + 2\sqrt{xy} + y}{x - y}$

50. $\frac{3}{\sqrt{6}} = \frac{3\sqrt{6}}{\sqrt{6}\sqrt{6}} = \frac{3\sqrt{6}}{6}$;

$\frac{2}{\sqrt{3}} = \frac{2\sqrt{3}}{\sqrt{3}\sqrt{3}} = \frac{2\sqrt{3}}{3}$;

$$\frac{3}{\sqrt{6}} - \frac{2}{\sqrt{3}} = \frac{3\sqrt{6}}{6} - \frac{2\sqrt{3}}{3} = \frac{3\sqrt{6}}{6} - \frac{4\sqrt{3}}{6} = \frac{3\sqrt{6} - 4\sqrt{3}}{6}$$

52. $\frac{1}{\sqrt{5}} = \frac{1\sqrt{5}}{\sqrt{5}\sqrt{5}} = \frac{\sqrt{5}}{5}$;

$$\sqrt{5} + \frac{\sqrt{1}}{\sqrt{5}} = \frac{\sqrt{5}}{1} + \frac{\sqrt{5}}{5} = \frac{5\sqrt{5}}{5} + \frac{\sqrt{5}}{5} = \frac{5\sqrt{5} + \sqrt{5}}{5} = \frac{6\sqrt{5}}{5}$$

54. $$\frac{3}{\sqrt{2x}} - \frac{1}{\sqrt{x}} = \frac{3}{\sqrt{2x}} - \frac{1 \cdot \sqrt{2}}{\sqrt{x} \cdot \sqrt{2}} = \frac{3}{\sqrt{2x}} - \frac{\sqrt{2}}{\sqrt{2x}} = \frac{3 - \sqrt{2}}{\sqrt{2x}}$$
$$= \frac{(3 - \sqrt{2})\sqrt{2x}}{\sqrt{2x}\,\sqrt{2x}} = \frac{3\sqrt{2x} - \sqrt{4x}}{2x} = \frac{3\sqrt{2x} - 2\sqrt{x}}{2x}$$

56. $$\frac{\sqrt{x^2 - 2}}{1} - \frac{x^2 + 1}{\sqrt{x^2 - 2}} = \frac{\sqrt{x^2 - 2}\,\sqrt{x^2 - 2}}{1\sqrt{x^2 - 2}} - \frac{x^2 + 1}{\sqrt{x^2 - 2}}$$
$$= \frac{x^2 - 2 - (x^2 + 1)}{\sqrt{x^2 - 2}}$$
$$= \frac{-3}{\sqrt{x^2 - 2}} = \frac{-3\sqrt{x^2 - 2}}{\sqrt{x^2 - 2}\,\sqrt{x^2 - 2}} = \frac{-3\sqrt{x^2 - 2}}{x^2 - 2}$$

58. $$\frac{x}{\sqrt{x^2 - 1}} + \frac{\sqrt{x^2 - 1}}{x} = \frac{x \cdot x}{x\sqrt{x^2 - 1}} + \frac{\sqrt{x^2 - 1}\,\sqrt{x^2 - 1}}{x\sqrt{x^2 - 1}}$$
$$= \frac{x^2 + x^2 - 1}{x\sqrt{x^2 - 1}} = \frac{2x^2 - 1}{x\sqrt{x^2 - 1}}$$
$$= \frac{(2x^2 - 1)\sqrt{x^2 - 1}}{x\sqrt{x^2 - 1}\,\sqrt{x^2 - 1}} = \frac{(2x^2 - 1)\sqrt{x^2 - 1}}{x(x^2 - 1)}$$

60. $$\frac{(\sqrt{3} + \sqrt{2})(\sqrt{3} - \sqrt{2})}{3(\sqrt{3} - \sqrt{2})} = \frac{3 - 2}{\sqrt{3}(\sqrt{3} - \sqrt{2})} = \frac{1}{\sqrt{3}(\sqrt{3} - \sqrt{2})} \text{ or } \frac{1}{3 - \sqrt{6}}$$

62. $$\frac{(4 - \sqrt{2y})(4 + \sqrt{2y})}{2(4 + \sqrt{2y})} = \frac{16 - 2y}{2(4 + \sqrt{2y})} = \frac{2(8 - y)}{2(4 + \sqrt{2y})} = \frac{8 - y}{4 + \sqrt{2y}}$$

64. $$\frac{(2\sqrt{x} + \sqrt{y})(2\sqrt{x} - \sqrt{y})}{\sqrt{xy}(2\sqrt{x} - \sqrt{y})} = \frac{4x - y}{\sqrt{xy}(2\sqrt{x} - \sqrt{y})} \text{ or } \frac{4x - y}{2x\sqrt{y} - y\sqrt{x}}$$

6

SECOND-DEGREE EQUATIONS AND INEQUALITIES

EXERCISE 6.1

2. $x + 3 = 0; \quad x - 4 = 0$
   $\{-3,4\}$

4. $x + 1 = 0; \quad 3x - 1 = 0$
   $3x = 1$
   $\left\{-1,\frac{1}{3}\right\}$

6. $x = 0; \quad 3x - 7 = 0$
   $3x = 7$
   $\left\{0,\frac{7}{3}\right\}$

8. $2x - 7 = 0; \quad x + 1 = 0$
   $2x = 7$
   $\left\{\frac{7}{2}, -1\right\}$

10. $x + 5 = 0; \quad 3x - 1 = 0$
    $3x = 1$
    $\left\{-5,\frac{1}{3}\right\}$

12. $3x - 4 = 0; \quad 2x + 5 = 0$
    $3x = 4 \quad 2x = -5$
    $\left\{\frac{4}{3}, \frac{-5}{2}\right\}$

14. $x(x + 5) = 0$
    $x = 0; \quad x + 5 = 0$
    $\{0,-5\}$

16. $3x^2 - 3x = 0$
    $3x(x - 1) = 0$
    $x = 0; \quad x - 1 = 0$
    $\{0,1\}$

18. $(x + 2)(x - 2) = 0$
    $x + 2 = 0; \quad x - 2 = 0$
    $\{-2,2\}$

20. $3(x^2 - 1) = 0$
    $3(x + 1)(x - 1) = 0$
    $x + 1 = 0; \quad x - 1 = 0$
    $\{-1,1\}$

22. $5x\left(5x - \frac{4}{5x}\right) = 5x(0)$

$25x^2 - 4 = 0$
$(5x + 2)(5x - 2) = 0$
$5x + 2 = 0; \quad 5x - 2 = 0$
$5x = -2; \quad 5x = 2$
$x = \frac{-2}{5}; \quad x = \frac{2}{5}$
$\left\{\frac{-2}{5}, \frac{2}{5}\right\}$

24. $25\left(\frac{9x^2}{25} - 1\right) = (25)0$

$9x^2 - 25 = 0$
$(3x + 5)(3x - 5) = 0$
$3x + 5 = 0; \quad 3x - 5 = 0$
$3x = -5; \quad 3x = 5$
$\left\{\frac{-5}{3}, \frac{5}{3}\right\}$

26. $(x + 2)(x + 3) = 0$
$x + 2 = 0; \quad x + 3 = 0$
$\{-2, -3\}$

28. $(x + 6)(x - 7) = 0$
$x + 6 = 0; \quad x - 7 = 0$
$\{-6, 7\}$

30. $12x^2 - 8x - 15 = 0$
$(2x - 3)(6x + 5) = 0$
$2x - 3 = 0; \quad 6x + 5 = 0$
$2x = 3; \quad 6x = -5$
$\left\{\frac{3}{2}, \frac{-5}{6}\right\}$

32. $2x^2 - 4x = x + 3$
$2x^2 - 5x - 3 = 0$
$(2x + 1)(x - 3) = 0$
$2x + 1 = 0; \quad x - 3 = 0$
$2x = -1$
$\left\{\frac{-1}{2}, 3\right\}$

34. $3x^2 + 2x = x^2 + 4x + 4$
$2x^2 - 2x - 4 = 0$
$2(x^2 - x - 2) = 0$
$2(x + 1)(x - 2) = 0$
$x + 1 = 0; \quad x - 2 = 0$
$\{-1, 2\}$

36. $x^2 + x = 4 - [x^2 + 4x + 4]$
$x^2 + x = 4 - x^2 - 4x - 4$
$2x^2 + 5x = 0$
$x(2x + 5) = 0$
$x = 0; \quad 2x + 5 = 0$
$\left\{0, \frac{-5}{2}\right\}$

38. $z^2 + 5z + 18 = 4 - 4z$
$z^2 + 9z + 14 = 0$
$(z + 2)(z + 7) = 0$
$z + 2 = 0; \quad z + 7 = 0$
$\{-2, -7\}$

40. $n^2 - n - 6 = 6$
$n^2 - n - 12 = 0$
$(n - 4)(n + 3) = 0$
$n - 4 = 0; \quad n + 3 = 0$
$\{-3, 4\}$

42.
$$2x^2 - x - 3 = 3$$
$$2x^2 - x - 6 = 0$$
$$(2x + 3)(x - 2) = 0$$
$$2x + 3 = 0; \quad x - 2 = 0$$
$$2x = -3; \quad x = 2$$
$$\left\{-\frac{3}{2}, 2\right\}$$

44.
$$3\left(2x - \frac{5}{3}\right) = 3\left(\frac{x^2}{3}\right)$$
$$6x - 5 = x^2$$
$$0 = x^2 - 6x + 5$$
$$0 = (x - 1)(x - 5)$$
$$x - 1 = 0; \quad x - 5 = 0$$
$$\{1,5\}$$

46.
$$4x\left(\frac{x}{4} - \frac{3}{4}\right) = 4x\left(\frac{1}{x}\right)$$
$$x^2 - 3x = 4$$
$$x^2 - 3x - 4 = 0$$
$$(x + 1)(x - 4) = 0$$
$$x + 1 = 0; \quad x - 4 = 0$$
$$\{-1,4\}$$

48.
$$[3x(3x + 1)]\frac{4}{3x} + [3x(3x + 1)]\frac{3}{(3x + 1)} + [3x(3x + 1)]2 = [3x(3x + 1)] \cdot 0$$
$$4(3x + 1) + (3x)3 + 6x(3x + 1) = 0$$
$$12x + 4 + 9x + 18x^2 + 6x = 0$$
$$18x^2 + 27x + 4 = 0$$
$$(3x + 4)(6x + 1) = 0$$
$$3x + 4 = 0; \quad 6x + 1 = 0$$
$$\left\{\frac{-4}{3}, \frac{-1}{6}\right\}$$

50.
$$[x - (-4)](x - 3) = 0$$
$$(x + 4)(x - 3) = 0$$
$$x^2 + x - 12 = 0$$

52.
$$x(x - 5) = 0$$
$$x^2 - 5x = 0$$

54.
$$\left[x - \left(-\frac{2}{3}\right)\right](x - 3) = 0$$
$$\left(x + \frac{2}{3}\right)(x - 3) = 0$$
$$x^2 + \frac{2}{3}x - 3x - 2 = 0$$
$$3\left(x^2 + \frac{2}{3}x - 3x - 2\right) = 3(0)$$
$$3x^2 + 2x - 9x - 6 = 0$$
$$3x^2 - 7x - 6 = 0$$

56.
$$\left(x - \frac{2}{3}\right)\left(x - \frac{1}{5}\right) = 0$$
$$x^2 - \frac{2}{3}x - \frac{1}{5}x + \frac{2}{15} = 0$$
$$15\left(x^2 - \frac{2}{3}x - \frac{1}{5}x + \frac{2}{15}\right) = 15(0)$$
$$15x^2 - 10x - 3x + 2 = 0$$
$$15x^2 - 13x + 2 = 0$$

58.
$$[x - (a + b)][x + (a + b)] = 0$$
$$[x - (a + b)] = 0; \quad [x + (a + b)] = 0$$
$$x = a + b; \quad x = -a - b$$
$$\{a + b, -a - b\}$$

60.
$$(x + 6b)(x - 2b) = 0$$
$$x + 6b = 0; \quad x - 2b = 0$$
$$\{-6b, 2b\}$$

62.
$$(x + 2a)(x - b) = 0$$
$$x + 2a = 0; \quad x - b = 0$$
$$\{-2a, b\}$$

EXERCISE 6.2

2. $x = 4; \quad x = -4$
$\{-4,4\}$

4. $x^2 = \frac{9}{4}$
$x = \frac{3}{2}; \quad x = \frac{-3}{2}$
$\left\{\frac{-3}{2}, \frac{3}{2}\right\}$

6. $x^2 = 5$
$x = \sqrt{5}; \quad x = -\sqrt{5}$
$\{-\sqrt{5}\ \sqrt{5}\}$

8. $3x^2 = 9$
$x^2 = 3$
$x = \sqrt{3}; \quad x = -\sqrt{3}$
$\{-\sqrt{3}\ \sqrt{3}\}$

10. $5\left(\frac{3x^2}{5}\right) = 5(3)$
$3x^2 = 15$
$x^2 = 5$
$x = \sqrt{5}; \quad x = -\sqrt{5}$
$\{-\sqrt{5}, \sqrt{5}\}$

12. $2\left(\frac{9x^2}{2}\right) = 2(50)$
$9x^2 = 100$
$x^2 = \frac{100}{9}$
$x = \frac{10}{3}; \quad x = \frac{-10}{3}$
$\left\{\frac{10}{3}, \frac{-10}{3}\right\}$

14. $x + 3 = 2; \quad x + 3 = -2$
$x = -1; \quad x = -5$
$\{-5,-1\}$

16. $3x + 1 = 5; \quad 3x + 1 = -5$
$3x = 4; \quad 3x = -6$
$\left\{-2, \frac{4}{3}\right\}$

18. $x - 5 = \sqrt{7}; \quad x - 5 = -\sqrt{7}$
$x = 5 + \sqrt{7}; \quad x = 5 - \sqrt{7}$
$\{5 + \sqrt{7}, 5 - \sqrt{7}\}$

20. $x + 3 = \sqrt{18}; \quad x + 3 = -\sqrt{18}$
$x = -3 + \sqrt{18}; \quad x = -3 - \sqrt{18}$
$x = -3 + 3\sqrt{2}; \quad x = -3 - 3\sqrt{2}$
$\{-3 - 3\sqrt{2}, -3 + 3\sqrt{2}\}$

22. $x + 3 = 6; \quad x + 3 = -6$
$x = 3; \quad x = -9$
$\{-9,3\}$

24. $3x + 4 = 4; \quad 3x + 4 = -4$
$3x = 0; \quad 3x = -8$
$x = 0 \quad x = \frac{-8}{3}$
$\left\{\frac{-8}{3}, 0\right\}$

26. $2x + 3 = 6; \quad 2x + 3 = -6$
$2x = 3; \quad 2x = -9$
$\left\{\frac{-9}{2}, \frac{3}{2}\right\}$

28. $5x + 3 = \sqrt{7}; \quad 5x + 3 = -\sqrt{7}$
$5x = \sqrt{7} - 3; \quad 5x = -\sqrt{7} - 3$
$\left\{\frac{\sqrt{7} - 3}{5}, \frac{-\sqrt{7} - 3}{5}\right\}$

30. $5x - 12 = \sqrt{24}$; $5x - 12 = -\sqrt{24}$

$5x = 2\sqrt{6} + 12$; $5x = -2\sqrt{6} + 12$

$\left\{\frac{2\sqrt{6} + 12}{5}, \frac{-2\sqrt{6} + 12}{5}\right\}$

32. $x^2 - x = 6$

$x^2 - x + \frac{1}{4} = 6 + \frac{1}{4}$

$\left(x - \frac{1}{2}\right)^2 = \frac{25}{4}$

$x - \frac{1}{2} = \frac{5}{2}$; $x - \frac{1}{2} = \frac{-5}{2}$

$x = 3$; $x = -2$

$\{-2,3\}$

34. $x^2 + 4x = -4$

$x^2 + 4x + 4 = -4 + 4$

$(x + 2)^2 = 0$

$x + 2 = 0$

$\{-2\}$

36. $x^2 - x = 20$

$x^2 - x + \frac{1}{4} = 20 + \frac{1}{4}$

$\left(x - \frac{1}{2}\right)^2 = \frac{81}{4}$

$x - \frac{1}{2} = \frac{9}{2}$; $x - \frac{1}{2} = \frac{-9}{2}$

$x = 5$ $x = -4$

$\{-4,5\}$

38. $x^2 + 3x = 1$

$x^2 + 3x + \frac{9}{4} = 1 + \frac{9}{4}$

$\left(x + \frac{3}{2}\right)^2 = \frac{13}{4}$

$x + \frac{3}{2} = \frac{\sqrt{13}}{2}$; $x + \frac{3}{2} = \frac{-\sqrt{13}}{2}$

$x = \frac{-3}{2} + \frac{\sqrt{13}}{2}$; $x = \frac{-3}{2} - \frac{\sqrt{13}}{2}$

$\left\{\frac{-3 + \sqrt{13}}{2}, \frac{-3 - \sqrt{13}}{2}\right\}$

40. $2x^2 + 5x = 6$

$x^2 + \frac{5}{2}x = 3$

$x^2 + \frac{5}{2}x + \frac{25}{16} = 3 + \frac{25}{16}$

$\left(x + \frac{5}{4}\right)^2 = \frac{73}{16}$

$x + \frac{5}{4} = \frac{\sqrt{73}}{4}$; $x + \frac{5}{4} = \frac{-\sqrt{73}}{4}$

$x = \frac{-5}{4} + \frac{\sqrt{73}}{4}$; $x = \frac{-5}{4} - \frac{\sqrt{73}}{4}$

$\left\{\frac{-5 + \sqrt{73}}{4}, \frac{-5 - \sqrt{73}}{4}\right\}$

42. $x^2 + \frac{1}{3}x = \frac{4}{3}$

$x^2 + \frac{1}{3}x + \frac{1}{36} = \frac{4}{3} + \frac{1}{36}$

$\left(x + \frac{1}{6}\right)^2 = \frac{49}{36}$

$x + \frac{1}{6} = \frac{7}{6}$; $x + \frac{1}{6} = \frac{-7}{6}$

$x = 1$; $x = \frac{-8}{6}$

$= \frac{-4}{3}$

$\left\{1, \frac{-4}{3}\right\}$

44.
$$x^2 - \frac{3}{4} = \frac{1}{2}x$$
$$x^2 - \frac{1}{2}x = \frac{3}{4}$$
$$x^2 - \frac{1}{2}x + \frac{1}{16} = \frac{3}{4} + \frac{1}{16}$$
$$\left(x - \frac{1}{4}\right)^2 = \frac{13}{16}$$
$$x - \frac{1}{4} = \frac{\sqrt{13}}{4};\ x - \frac{1}{4} = \frac{-\sqrt{13}}{4}$$
$$x = \frac{1}{4} + \frac{\sqrt{13}}{4};\ x = \frac{1}{4} - \frac{\sqrt{13}}{4}$$
$$\left\{\frac{1 + \sqrt{13}}{4}; \frac{1 - \sqrt{13}}{4}\right\}$$

46.
$$x^2 - \frac{5}{2}x = 4$$
$$x^2 - \frac{5}{2}x + \frac{25}{16} = 4 + \frac{25}{16}$$
$$\left(x - \frac{5}{4}\right)^2 = \frac{89}{16}$$
$$x - \frac{5}{4} = \frac{\sqrt{89}}{4};\ x - \frac{5}{4} = \frac{-\sqrt{89}}{4}$$
$$x = \frac{5}{4} + \frac{\sqrt{89}}{4};\ x = \frac{5}{4} - \frac{\sqrt{89}}{4}$$
$$\left\{\frac{5 + \sqrt{89}}{4}, \frac{5 - \sqrt{89}}{4}\right\}$$

48.
$$x^2 - \frac{7}{3}x = 1$$
$$x^2 - \frac{7}{3}x + \frac{49}{36} = 1 + \frac{49}{36}$$
$$\left(x - \frac{7}{6}\right)^2 = \frac{85}{36}$$
$$x - \frac{7}{6} = \frac{\sqrt{85}}{6};\ x - \frac{7}{6} = \frac{-\sqrt{85}}{6}$$
$$x = \frac{7}{6} + \frac{\sqrt{85}}{6};\ x = \frac{7}{6} - \frac{\sqrt{85}}{6}$$
$$\left\{\frac{7 + \sqrt{85}}{6}, \frac{7 - \sqrt{85}}{6}\right\}$$

50.
$$x^2 = 2a$$
$$x = \sqrt{2a};\ x = -\sqrt{2a}$$
$$\{\sqrt{2a}, -\sqrt{2a}\}$$

52.
$$\frac{c}{1}\left(\frac{bx^2}{c} - a\right) = \left(\frac{c}{1}\right)0$$
$$bx^2 - ac = 0$$
$$bx^2 = ac$$
$$x^2 = \frac{ac}{b}$$
$$x = \sqrt{\frac{ac}{b}} \qquad x = -\sqrt{\frac{ac}{b}}$$
$$= \frac{\sqrt{abc}}{b}; \qquad = -\frac{\sqrt{abc}}{b}$$
$$\left\{\frac{\sqrt{abc}}{b}, \frac{-\sqrt{abc}}{b}\right\}$$

54.
$$(x + a)^2 = 36$$
$$x + a = 6;\ x + a = -6$$
$$x = -a + 6;\ x = -a - 6$$
$$\{-a + 6, -a - 6\}$$

56.
$$ax - b = 5;\ ax - b = -5$$
$$ax = b + 5;\ ax = b - 5$$
$$\left\{\frac{b + 5}{a}, \frac{b - 5}{a}\right\}$$

58.
$$r^2 + S = h$$
$$r^2 = h - S$$
$$r = \sqrt{h - S};\ r = -\sqrt{h - S}$$
$$\{\sqrt{h - S}, -\sqrt{h - S}\}$$

60.
$$2(s) = 2\left(\frac{1}{2}gt^2\right) + 2c$$
$$2s = gt^2 + 2c$$
$$gt^2 + 2c = 2s$$
$$t^2 = \frac{2s - 2c}{g}$$
$$t = \sqrt{\frac{2s - 2c}{g}}; \quad t = \sqrt{\frac{2s - 2c}{g}}$$
$$= \frac{\sqrt{2gs - 2gc}}{g}; \quad = \frac{-\sqrt{2gs - 2gc}}{g}$$
$$\left\{\frac{\sqrt{2gs - 2gc}}{g}, \frac{-\sqrt{2gs - 2gc}}{g}\right\}$$

EXERCISE 6.3

2. $\sqrt{-1}\sqrt{9} = 3i$

4. $\sqrt{-1 \cdot 25 \cdot 2} = \sqrt{-1}\sqrt{25}\sqrt{2}$
$= i \cdot 5 \cdot \sqrt{2}$
$= 5i\sqrt{2}$

6. $4\sqrt{-1 \cdot 9 \cdot 2} = 4\sqrt{-1}\sqrt{9}\sqrt{2}$
$= 4 \cdot i \cdot 3\sqrt{2}$
$= 12i\sqrt{2}$

8. $2\sqrt{-1 \cdot 4 \cdot 10} = 2 \cdot \sqrt{-1} \cdot 2 \cdot \sqrt{10}$
$= 4i\sqrt{10}$

10. $7\sqrt{-1 \cdot 81} = 7 \cdot \sqrt{-1} \cdot \sqrt{81}$
$= 7 \cdot i \cdot 9$
$= 63i$

12. $-3\sqrt{-1 \cdot 25 \cdot 3}$
$= -3 \cdot \sqrt{-1} \cdot \sqrt{25} \cdot \sqrt{3}$
$= -3 \cdot i \cdot 5 \cdot \sqrt{3}$
$= -15i\sqrt{3}$

14. $5 - 3i$

16. $5\sqrt{-1 \cdot 4 \cdot 3} - 1$
$= 5\sqrt{-1}\sqrt{4}\sqrt{3} - 1$
$= 5i \cdot 2\sqrt{3} - 1$
$= 10i\sqrt{3} - 1$

18. $\sqrt{4 \cdot 5} - \sqrt{-1 \cdot 4 \cdot 5}$
$= \sqrt{4}\sqrt{5} - \sqrt{-1}\sqrt{4}\sqrt{5}$
$= 2\sqrt{5} - i \cdot 2 \cdot \sqrt{5}$
$= 2\sqrt{5} - 2i\sqrt{5}$

20. $(2 + 3) + [-1 + (-2)]i$
$= 5 + (-3]i$
$= 5 - 3i$

22. $(2 + i) + (-4 + 2i)$
$= [2 + (-4)] + (1 + 2)i$
$= -2 + 3i$

24. $(2 - 6i) + (-3 + 0i)$
$= [2 + (-3)] + (-6 + 0)i$
$= -1 - 6i$

26. $4 - 5i - 12i + 15i^2$
$= 4 - 17i + 15(-1)$
$= -11 - 17i$

28. $-6 - 2i + 9i + 3i^2$
$= -6 + 7i + 3(-1)$
$= -9 + 7i$

30. $-14 - 6i - 21i - 9i^2$
$= -14 - 27i - 9(-1)$
$= -5 - 27i$

32. $(2 + 3i)(2 + 3i)$
$= 4 + 6i + 6i + 9i^2$
$= 4 + 12i + 9(-1)$
$= -5 + 12i$

34. $1 - 4i^2 = 1 - 4(-1)$
$= 5$

36. $\dfrac{-2 \cdot i}{5i \cdot i} = \dfrac{-2i}{5i^2} = \dfrac{-2i}{5(-1)} = \dfrac{2}{5}i$

38. $\dfrac{(4 + 2i)i}{3i \cdot i} = \dfrac{4i + 2i^2}{3i^2}$
$= \dfrac{4i + 2(-1)}{3(-1)}$
$= \dfrac{-2 + 4i}{-3}$
$= \dfrac{2}{3} - \dfrac{4}{3}i$

40. $\dfrac{-3(2 - i)}{(2 + i)(2 - i)} = \dfrac{-6 + 3i}{4 - i^2}$
$= \dfrac{-6 + 3i}{4 - (-1)}$
$= \dfrac{-6 + 3i}{5}$
$= -\dfrac{6}{5} + \dfrac{3}{5}i$

42. $\dfrac{(3 - i)(1 - i)}{(1 + i)(1 - i)}$
$= \dfrac{3 - 4i + i^2}{1 - i^2}$
$= \dfrac{3 - 4i + (-1)}{1 - (-1)}$
$= \dfrac{2 - 4i}{2} = 1 - 2i$

44. $\dfrac{(6 + i)(2 + 5i)}{(2 - 5i)(2 + 5i)}$
$= \dfrac{12 + 32i + 5i^2}{4 - 25i^2}$
$= \dfrac{12 + 32i + 5(-1)}{4 - 25(-1)}$
$= \dfrac{7 + 32i}{29} = \dfrac{7}{29} + \dfrac{32}{29}i$

46. $\dfrac{(-4 - 3i)(2 - 7i)}{(2 + 7i)(2 - 7i)}$
$= \dfrac{-8 + 22i + 21i^2}{4 - 49i^2}$
$= \dfrac{-8 + 22i + 21(-1)}{4 - 49(-1)}$
$= \dfrac{-29 + 22i}{53} = -\dfrac{29}{53} + \dfrac{22}{53}i$

48. $i\sqrt{9}(3 + i\sqrt{16}) = 3i(3 + 4i)$
$= 9i + 12i^2$
$= 9i - 12$
$= -12 + 9i$

50. $(4 - i\sqrt{2})(3 + i\sqrt{2})$
$= 12 + 4i\sqrt{2} - 3i\sqrt{2} - 2i^2$
$= 12 + i\sqrt{2} + 2$
$= 14 + i\sqrt{2}$

52. $\dfrac{-1}{i\sqrt{25}} = \dfrac{(-1) \cdot i}{(5i) \cdot i} = \dfrac{-i}{5i^2}$
$= \dfrac{-i}{-5} = \dfrac{1}{5}i$

54. $\frac{1 + i\sqrt{2}}{3 - i\sqrt{3}} = \frac{(1 + i\sqrt{2})(3 + i\sqrt{3})}{(3 - i\sqrt{3})(3 + i\sqrt{3})} = \frac{3 + i\sqrt{3} + 3i\sqrt{2} + \sqrt{6}i^2}{9 - 3i^2}$

$= \frac{3 + (\sqrt{3} + 3\sqrt{2})i - \sqrt{6}}{9 + 3}$

$= \frac{3 - \sqrt{6}}{12} + \frac{\sqrt{3} + 3\sqrt{2}}{12}i$

56. $x^2 = -25$
$x = \sqrt{-25}$; $x = -\sqrt{-25}$
$x = i\sqrt{25}$; $x = -i\sqrt{25}$
$\{5i, -5i\}$

58. $x^2 = -24$
$x = \sqrt{-24}$; $x = -\sqrt{-24}$
$x = i\sqrt{24}$; $x = -i\sqrt{24}$
$\{2i\sqrt{6}, -2i\sqrt{6}\}$

60. $x^2 + x = -1$

$x^2 + x + \frac{1}{4} = -1 + \frac{1}{4}$

$\left(x + \frac{1}{2}\right)^2 = -\frac{3}{4}$

$x + \frac{1}{2} = \sqrt{\frac{-3}{4}}$; $x + \frac{1}{2} = -\sqrt{\frac{-3}{4}}$

$x + \frac{1}{2} = i\frac{\sqrt{3}}{2}$; $x + \frac{1}{2} = -i\frac{\sqrt{3}}{2}$

$x = -\frac{1}{2} + i\frac{\sqrt{3}}{2}$; $x = -\frac{1}{2} - i\frac{\sqrt{3}}{2}$

$-\frac{1}{2} + i\frac{\sqrt{3}}{2}, -\frac{1}{2} - i\frac{\sqrt{3}}{2}$

62. $x^2 - 4x = -5$

$x^2 - 4x + 4 = -5 + 4$

$(x - 2)^2 = -1$

$x - 2 = \sqrt{-1}$; $x - 2 = -\sqrt{-1}$
$x = 2 + i$; $x = 2 - i$
$\{2 + i, 2 - i\}$

64. Assuming that properties of exponents that hold for real number bases also hold for bases that are imaginary numbers:

a. $i^{-1} = \frac{1}{i} = \frac{1 \cdot i}{i \cdot i} = \frac{i}{-1} = -i$

b. $i^{-2} = \frac{1}{i^2} = \frac{1}{-1} = -1$

c. $i^{-3} = (i^{-1})(i^{-2})$
$= (-i)(-1) = i$

66. $2(2 - i)^2 - (2 - i) + 2 = 2(4 - 4i + i^2) - 2 + i + 2$
$= 8 - 8i + 2i^2 - 2 + i + 2$
$= 8 - 8i - 2 - 2 + i + 2$
$= 6 - 7i$

68. For $\sqrt{x + 3}$ to be real,
$x + 3 \geqslant 0$ or $x \geqslant -3$.
For $\sqrt{x + 3}$ to be imaginary,
$x + 3 < 0$ or $x < -3$.

EXERCISE 6.4

2. $a = 1,\ b = -4,\ c = 4$

$$x = \frac{-(-4) \pm \sqrt{(-4)^2 - 4(1)(4)}}{2(1)}$$

$$= \frac{4 \pm 0}{2}; \qquad \{2\}$$

4. $y^2 - 5y - 6 = 0;\quad a = 1,\ b = -5,\ c = -6$

$$y = \frac{-(-5) \pm \sqrt{(-5)^2 - 4(1)(-6)}}{2(1)} = \frac{5 \pm \sqrt{49}}{2}$$

$$y = \frac{5 + 7}{2},\quad y = \frac{5 - 7}{2}; \qquad \{6, -1\}$$

6. $2z^2 - 7z + 6 = 0;\quad a = 2,\ b = -7,\ c = 6$

$$z = \frac{-(-7) \pm \sqrt{(-7)^2 - 4(2)(6)}}{2(2)} = \frac{7 \pm \sqrt{1}}{4}$$

$$z = \frac{7 + 1}{4},\quad z = \frac{7 - 1}{4}; \qquad \left\{2, \frac{3}{2}\right\}$$

8. $2x^2 - x + 1 = 0;\quad a = 2,\ b = -1,\ c = 1$

$$x = \frac{-(-1) \pm \sqrt{(-1)^2 - 4(2)(1)}}{2(2)} = \frac{1 \pm \sqrt{-7}}{4}$$

$$\left\{\frac{1 + i\sqrt{7}}{4}, \frac{1 - i\sqrt{7}}{4}\right\}$$

10. $6z^2 - 13z - 5 = 0; \quad a = 6, b = -13, c = -5$

$$z = \frac{-(-13) \pm \sqrt{(-13)^2 - 4(6)(-5)}}{2(6)} = \frac{13 \pm \sqrt{289}}{12}$$

$$z = \frac{13 + 17}{12}; \quad z = \frac{13 - 17}{12}$$

$$z = \frac{30}{12}; \quad z = \frac{-4}{12}; \quad \left\{\frac{5}{2}, \frac{-1}{3}\right\}$$

12. $a = 1, b = 3, c = 0$

$$y = \frac{-3 \pm \sqrt{3^2 - 4(1)(0)}}{2(1)} = \frac{-3 \pm 3}{2}$$

$$y = \frac{-3 + 3}{2}; \quad y = \frac{-3 - 3}{2}; \quad \{0, -3\}$$

14. $a = 2; b = 0, c = 1$

$$z = \frac{0 \pm \sqrt{0^2 - 4(2)(1)}}{2(2)} = \frac{\pm\sqrt{-8}}{4} = \frac{\pm 2i\sqrt{2}}{4}; \quad \left\{\frac{i\sqrt{2}}{2}, \frac{-i\sqrt{2}}{2}\right\}$$

16. $x^2 + 2x + 5 = 0; \quad a = 1, b = 2, c = 5$

$$x = \frac{-2 \pm \sqrt{2^2 - 4(1)(5)}}{2(1)} = \frac{-2 \pm \sqrt{-16}}{2} = \frac{-2 \pm 4i}{2}$$

$$= -1 \pm 2i; \quad \{-1 + 2i, -1 - 2i\}$$

18. $2z(z - 2) = 3$

$2z^2 - 4z - 3 = 0; \quad a = 2, b = -4, c = -3$

$$z = \frac{-(-4) \pm \sqrt{(-4)^2 - 4(2)(-3)}}{2(2)} = \frac{4 \pm \sqrt{40}}{4}$$

$$= \frac{4 \pm 2\sqrt{10}}{4} = \frac{2(2 \pm \sqrt{10})}{2 \cdot 2} = \frac{2 \pm \sqrt{10}}{2}$$

$$\left\{\frac{2 + \sqrt{10}}{2}, \frac{2 - \sqrt{10}}{2}\right\}$$

20. $2x(x - 1) = x + 1$

$2x^2 - 2x = x + 1$

$2x^2 - 3x - 1 = 0; \quad a = 2, b = -3, c = -1$

$$x = \frac{-(-3) \pm \sqrt{(-3)^2 - 4(2)(-1)}}{2(2)} = \frac{3 \pm \sqrt{17}}{4}$$

$$\left\{\frac{3 + \sqrt{17}}{4}, \frac{3 - \sqrt{17}}{4}\right\}$$

22. $3z^2 + z + 2 = 0;\quad a = 3,\ b = 1,\ c = 2$

$$z = \frac{-1 \pm \sqrt{1^2 - 4(3)(2)}}{2(3)} = \frac{-1 \pm \sqrt{-23}}{6} = \frac{-1 \pm i\sqrt{23}}{6};$$

$$\left\{\frac{-1}{6} + \frac{\sqrt{23}}{6}i,\ \frac{-1}{6} - \frac{\sqrt{23}}{6}i\right\}$$

24. $y(y - 1) = -1$

$y^2 - y + 1 = 0;\quad a = 1,\ b = -1,\ c = 1$

$$y = \frac{-(-1) \pm \sqrt{(-1)^2 - 4(1)(1)}}{2(1)} = \frac{1 \pm \sqrt{-3}}{2} = \frac{1 \pm i\sqrt{3}}{2};$$

$$\left\{\frac{1}{2} + \frac{\sqrt{3}}{2}i, \frac{1}{2} - \frac{\sqrt{3}}{2}i\right\}$$

26. $a = 1,\ b = -2,\ c = -3$
$b^2 - 4ac = 4 - (-12) = 16$
$b^2 - 4ac > 0$; therefore, the solutions are real and unequal.

28. $a = 2,\ b = 3,\ c = 7$
$b^2 - 4ac = 9 - 56 = -47$
$b^2 - 4ac < 0$; therefore, the solutions are imaginary and unequal.

30. $2z^2 - z - 12 = 0;\quad a = 2,\ b = -1,\ c = -12$
$b^2 - 4ac = 1 - 4(-24) = 97$
$b^2 - 4ac > 0$; therefore, the solutions are real and unequal.

32. $a = 2,\ b = -k,\ c = 3$

$$x = \frac{-(-k) \pm \sqrt{(-k)^2 - 4(2)(3)}}{2(2)} = \frac{k \pm \sqrt{k^2 - 24}}{4}$$

34. $x^2 + 2x + (c + 3) = 0;\quad a = 1,\ b = 2,\ c = c + 3$

$$x = \frac{-2 \pm \sqrt{2^2 - 4(1)(c + 3)}}{2(1)} = \frac{-2 \pm \sqrt{-4c - 8}}{2}$$

$$= \frac{-2 \pm 2\sqrt{-c - 2}}{2} = -1 \pm \sqrt{-c - 2}$$

36. $a = 2,\ b = -3,\ c = 2y$

$$x = \frac{-(-3) \pm \sqrt{(-3)^2 - 4(2)(2y)}}{2(2)} = \frac{3 \pm \sqrt{9 - 16y}}{4}$$

38. $x^2 + (-3y)x + (y^2 - 3) = 0;\quad a = 1,\ b = -3y,\ c = y^2 - 3$

$$x = \frac{-(-3y) \pm \sqrt{(-3y)^2 - 4(1)(y^2 - 3)}}{2(1)} = \frac{3y \pm \sqrt{5y^2 + 12}}{2}$$

40. $y^2 + (-3x)y + (x^2 - 3) = 0$; $a = 1$, $b = -3x$, $c = x^2 - 3$

$$y = \frac{-(-3x) \pm \sqrt{(-3x)^2 - 4(1)(x^2 - 3)}}{2(1)} = \frac{3x \pm \sqrt{5x^2 + 12}}{2}$$

42. $a = 1$, $b = -k$, $c = 9$.

There will be one solution if $b^2 - 4ac = 0$. Hence, $k^2 - 36 = 0$; $k = 6$, $k = -6$

44. $x^2 - x + (k - 2) = 0$, $a = 1$, $b = -1$, $c = k - 2$

The solutions will be imaginary (not real numbers) if $b^2 - 4ac < 0$. Hence,

$$1 - 4(1)(k - 2) < 0$$
$$1 - 4k + 8 < 0$$
$$-4k < -9$$
$$k > \frac{9}{4}$$

46. Since $r_1 = \dfrac{-b + \sqrt{b^2 - 4ac}}{2a}$ is a solution of $ax^2 + bx + c = 0$, then $ar_1{}^2 + br_1 + c = 0$ or $ar_1{}^2 + br_1 = -c$. From the given equation in the exercise, $ar_1{}^2 + br_1 = -y$; therefore, $-y = -c$, or $y = c$.

EXERCISE 6.5

2. A positive integer: x
The next consecutive positive integer: $x + 1$

$$x(x + 1) = 56$$
$$x^2 + x - 56 = 0$$
$$(x + 8)(x - 7) = 0; \qquad \{-8, 7\}$$

-8 is not a positive integer. Hence, the two positive integers are 7 and 8.

4. A positive integer: x
The next consecutive positive integer: $x + 1$

$$x^2 + (x + 1)^2 = 61$$
$$x^2 + x^2 + 2x + 1 = 61$$
$$2x^2 + 2x - 60 = 0$$
$$x^2 + x - 30 = 0$$
$$(x + 6)(x - 5) = 0; \qquad \{-6, 5\}$$

-6 is not a positive integer. Hence, the two positive integers are 5 and 6.

6. The width of the plate: x
The length of the plate: $2x + 2$

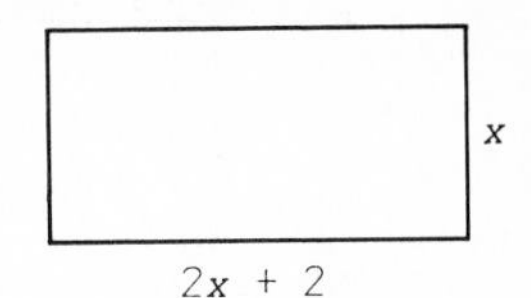

The area of a rectangle is equal to the product of its length and width.

$$x(2x + 2) = 40$$
$$2x^2 + 2x - 40 = 0$$
$$x^2 + x - 20 = 0$$
$$(x + 5)(x - 4) = 0$$
$$x = -5; \quad x = 4$$

Since a dimension of a geometric figure cannot be negative, -5 is rejected. Then when $x = 4$, $2x + 2 = 10$. The dimensions are 4 centimeters by 10 centimeters.

8. The number: n

The reciprocal of the number: $\frac{1}{n}$

$$n - \frac{2}{n} = \frac{17}{3}$$
$$3n\left(n - \frac{2}{n}\right) = 3n\left(\frac{17}{3}\right)$$
$$3n^2 - 6 = 17n$$
$$3n^2 - 17n - 6 = 0$$
$$(3n + 1)(n - 6) = 0$$
$$n = -\frac{1}{3}; \quad n = 6$$

The number is either $-\frac{1}{3}$ or 6.

10. Time it will take the ball to return to the ground: t

At the ground level, $h = 0$; hence,

$$0 = 64t - 16t^2$$
$$0 = 16t(4 - t)$$
$$t = 0; \quad t = 4.$$

The ball will return to the ground after 4 seconds.

12. Time it will take a body to fall 150 feet from rest: t

Since the body starts from rest, $v_0 = 0$; hence, with $s = 150$ and $g = 32$,

$$150 = 0 \cdot t + \frac{1}{2}(32)t^2$$
$$150 = 16t^2$$
$$t = \frac{\pm\sqrt{150}}{4} = \frac{\pm5\sqrt{6}}{4}.$$

Because $t = \frac{-5\sqrt{6}}{4}$ is not meaningful in this problem, we take $t = \frac{5\sqrt{6}}{4}$. It will take $\frac{5\sqrt{6}}{4}$ seconds.

14. The denominator: x
The numerator: $x - 2$

$$\frac{x-2}{x} + \frac{3}{1}\left(\frac{1}{\frac{x-2}{x}}\right) = \frac{28}{5}$$

$$\frac{x-2}{x} + \frac{3}{1}\left(\frac{x}{x-2}\right) = \frac{28}{5}$$

$$5x(x-2)\left[\frac{x-2}{x}\right] + 5x(x-2)\left[\frac{3x}{x-2}\right] = 5x(x-2)\left[\frac{28}{5}\right]$$

$$5(x-2)^2 + 15x^2 = 28x(x-2)$$
$$5x^2 - 20x + 20 + 15x^2 = 28x^2 - 56x$$
$$-8x^2 + 36x + 20 = 0$$
$$2x^2 - 9x - 5 = 0$$
$$(2x+1)(x-5) = 0$$

$x = \frac{-1}{2}$ and $x - 2 = \frac{-5}{2}$ or $x = 5$ and $x - 2 = 3$

The numerator and denominator are $\frac{-5}{2}$ and $\frac{-1}{2}$, or 3 and 5.

16. The altitude: x
The base: $x + 13$

The area of a triangle is equal to one-half the product of the base and altitude.

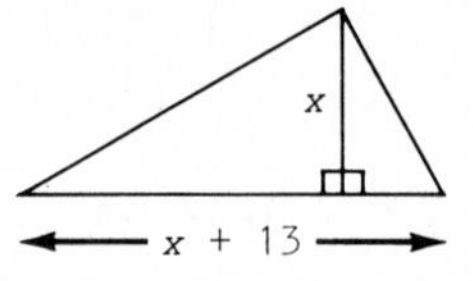

$$\frac{1}{2}(x)(x+13) = 70$$

$$2\left[\frac{1}{2}x(x+13)\right] = 2[70]$$

$$x(x+13) = 140$$
$$x^2 + 13x - 140 = 0$$
$$(x+20)(x-7) = 0; \qquad \{-20,7\}$$

Since -20 is not meaningful as a length, we have $x = 7$ and $x + 13 = 20$. The altitude and the base are 7 and 20 centimeters, respectively.

18. A positive integer: x
The next consecutive positive integer: $x + 1$

$$(x + 1)^3 - x^3 = 919$$
$$x^3 + 3x^2 + 3x + 1 - x^3 = 919$$
$$3x^2 + 3x - 918 = 0$$
$$\frac{1}{3}(3x^2 + 3x - 918) = \frac{1}{3}(0)$$
$$x^2 + x - 306 = 0$$
$$(x + 18)(x - 17) = 0; \qquad \{-18,17\}$$

Since -18 is not positive, we take $x = 17$ and $x + 1 = 18$. The integers are 17 and 18.

20. Number of consecutive natural numbers: n

$$\frac{n}{2}(n + 1) = 406$$
$$2\left[\frac{n}{2}(n + 1)\right] = 2[406]$$
$$n(n + 1) = 812$$
$$n^2 + n - 812 = 0$$
$$(n + 29)(n - 28) = 0; \qquad \{-29,28\}$$

Since -29 is not a natural number, we take $n = 28$. The number of consecutive natural numbers is 28.

22. Number of hours skiploader B takes for the job alone: x
Number of hours skiploader A takes for the job alone: $x - 1.5$

The part skiploader B would do alone in 1 hour: $\frac{1}{x}$

The part skiploader A would do alone in 1 hour: $\frac{1}{x - 1.5}$

Since they do the whole job working together for 1 hour, we have

$$\frac{1}{x} + \frac{1}{x - 1.5} = 1.$$

$$x(x - 1.5) \cdot \frac{1}{x} + x(x - 1.5) \cdot \frac{1}{x - 1.5} = x(x - 1.5) \cdot 1$$
$$(x - 1.5) + x = x(x - 1.5)$$
$$x - 1.5 + x = x^2 - 1.5x$$
$$0 = x^2 - 3.5x + 1.5$$
$$10x^2 - 35x + 15 = 0$$
$$2x^2 - 7x + 3 = 0$$
$$(2x - 1)(x - 3) = 0; \qquad \left\{\frac{1}{2}, 3\right\}$$

If $\frac{1}{2}$ is accepted as a solution, then $x - 1.5 = -1$. That is, skiploader *A* could do the job alone in -1 hours. Since this makes no sense, we reject this solution of the equation as a solution to the problem. Hence, skiploader *B* could do the job alone in 3 hours.

24. The width of the piece of metal: x
The length of the piece of metal: $x + 2$

A drawing is helpful.
The volume (160 cu cm) of such a box is equal to the product of its three dimensions.

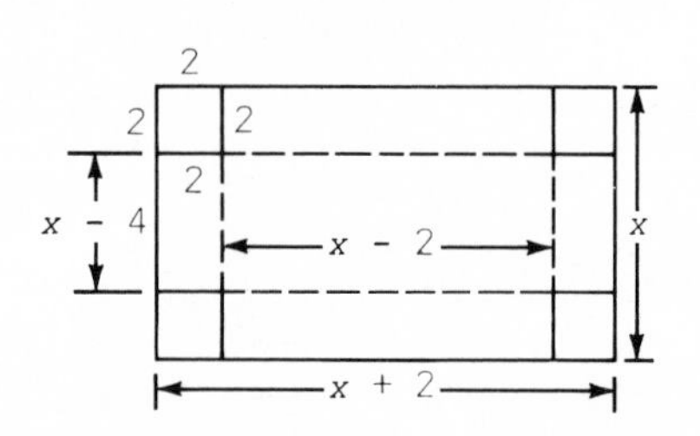

$$2(x - 2)(x - 4) = 160$$
$$2x^2 - 12x + 16 = 160$$
$$\frac{1}{2}(2x^2 - 12x + 16) = \frac{1}{2}(160)$$
$$x^2 - 6x + 8 = 80$$
$$x^2 - 6x - 72 = 0$$
$$(x - 12)(x + 6) = 0$$
$$x = 12; \quad x = -6$$

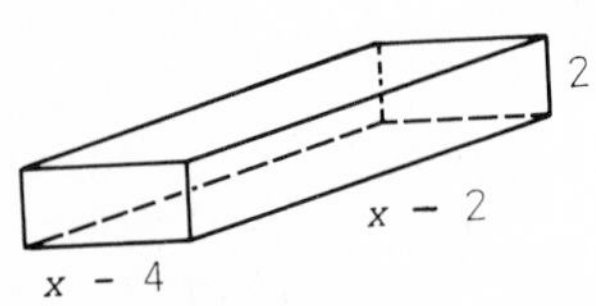

Since -6 is a negative number, we reject it as a length. Using x equal to 12, the dimensions in centimeters are $x - 2 = 10$, $x - 4 = 8$, and 2.

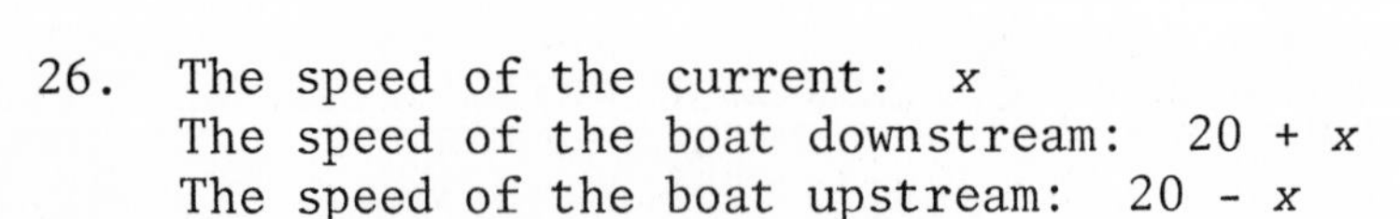

26. The speed of the current: x
The speed of the boat downstream: $20 + x$
The speed of the boat upstream: $20 - x$

	d	r	$t = d/r$
downstream	75	$20 + x$	$\frac{75}{20 + x}$
upstream	90	$20 - x$	$\frac{90}{20 - x}$

$$\begin{bmatrix}\text{time}\\ \text{upstream}\end{bmatrix} = \begin{bmatrix}\text{time}\\ \text{downstream}\end{bmatrix} + 3$$

$$\frac{90}{20 - x} = \frac{75}{20 + x} + 3$$

$$\begin{aligned}
(20 + x)(20 - x)\left[\frac{90}{20 - x}\right] &= (20 + x)(20 - x)\left[\frac{75}{20 + x} + 3\right]\\
90(20 + x) &= 75(20 - x) + 3(20 + x)(20 - x)\\
1800 + 90x &= 1500 - 75x + 1200 - 3x^2\\
3x^2 + 165x - 900 &= 0\\
\frac{1}{3}(3x^2 + 165x - 900) &= \frac{1}{3}(0)\\
x^2 + 55x - 330 &= 0\\
(x + 60)(x - 5) &= 0; \qquad \{-60,5\}
\end{aligned}$$

$x = -60$ is not meaningful, so it is rejected. We take $x = 5$. The speed of the current is 5 mph.

28. Rate in light traffic: x
Rate in heavy traffic: $x - 20$

	d	r	$t = d/r$
light traffic	40	x	$\frac{40}{x}$
heavy traffic	10	$x - 20$	$\frac{10}{x - 20}$

$$\begin{bmatrix}\text{time in light}\\ \text{traffic (in hours)}\end{bmatrix} + \begin{bmatrix}\text{time in heavy}\\ \text{traffic (in hours)}\end{bmatrix} = 1\frac{1}{2}$$

$$\frac{40}{x} + \frac{10}{x - 20} = \frac{3}{2}$$

$$\begin{aligned}
2x(x - 20)\left[\frac{40}{x} + \frac{10}{x - 20}\right] &= 2x(x - 20)\left[\frac{3}{2}\right]\\
80(x - 20) + 20x &= 3x(x - 20)\\
80x - 1600 + 20x &= 3x^2 - 60x\\
-3x^2 + 160x - 1600 &= 0\\
3x^2 - 160x + 1600 &= 0\\
(3x - 40)(x - 40) &= 0; \qquad \left\{\frac{40}{3}, 40\right\}
\end{aligned}$$

If x is taken as $\frac{40}{3}$, then $x - 20 = \frac{40}{3} - 20 = \frac{-20}{3}$, which is not meaningful. Hence, $x = 40$ and $x - 20 = 20$. The rates are 40 mph in light traffic and 20 mph in heavy traffic.

EXERCISE 6.6

2. $\sqrt{x} = 5$
$x = 25$

Check: $\sqrt{25} = 5$? Yes
$\{25\}$

4. $y - 3 = 25$
$y = 28$

Check: $\sqrt{28 - 3} = 5$? Yes
$\{28\}$

6.
$$(2z - 3)^2 = 7z - 3$$
$$4z^2 - 12z + 9 = 7z - 3$$
$$4z^2 - 19z + 12 = 0$$
$$(4z - 3)(z - 4) = 0$$
$$z = \frac{3}{4}, \quad z = 4$$

Check: $2\left(\frac{3}{4}\right) - 3 = \sqrt{7\left(\frac{3}{4}\right) - 3}$? No

$2(4) - 3 = \sqrt{7(4) - 3}$? Yes

$\{4\}$

8.
$$(4s + 5)^2 = 3s + 4$$
$$16s^2 + 40s + 25 = 3s + 4$$
$$16s^2 + 37s + 21 = 0$$
$$(16s + 21)(s + 1) = 0$$
$$s = \frac{-21}{16}, \quad s = -1$$

Check: $4\left(\frac{-21}{16}\right) + 5 = \sqrt{3\left(\frac{-21}{16}\right) + 4}$?

$\frac{-21}{4} + \frac{20}{4} = \sqrt{\frac{-63}{16} + \frac{64}{16}}$?

$-\frac{1}{4} = \sqrt{\frac{1}{16}}$? No

$4(-1) + 5 = \sqrt{3(-1) + 4}$?

$\{-1\}$ $\quad 1 = \sqrt{1}$? Yes

10.
$$16(x - 4) = x^2$$
$$16x - 64 = x^2$$
$$-x^2 + 16x - 64 = 0$$
$$x^2 - 16x + 64 = 0$$
$$(x - 8)(x - 8) = 0$$
$$x = 8$$

Check: $4\sqrt{8 - 4} = 8$?

$4\sqrt{4} = 8$? Yes

$\{8\}$

12.
$$4t + 1 = 6t - 3$$
$$-2t + 1 = -3$$
$$-2t = -4$$
$$t = 2$$

Check: $\sqrt{4(2) + 1} = \sqrt{6(2) - 3}$?

$\sqrt{9} = \sqrt{9}$? Yes

$\{2\}$

14.
$$x(x - 5) = 36$$
$$x^2 - 5x - 36 = 0$$
$$(x - 9)(x + 4) = 0$$
$$x = 9; \quad x = -4$$

Check: $\sqrt{9}\sqrt{9 - 5} = 6$? Yes

$\sqrt{-4}\sqrt{-4 - 5} = 6$? No

$\{9\}$

16. $x = (-4)^3$
$x = -64$

Check: $\sqrt[3]{-64} = -4$? Yes
$\{-64\}$

18. $x - 1 = 81$
$x = 82$

Check: $\sqrt[4]{82 - 1} = 3$? Yes
$\{82\}$

20. $\sqrt{1 + 16y} = 5 - 4\sqrt{y}$
$1 + 16y = (5 - 4\sqrt{y})^2$
$1 + 16y = 25 - 40\sqrt{y} + 16y$
$-24 = -40\sqrt{y}$
$\sqrt{y} = \frac{24}{40} = \frac{3}{5}$; $y = \frac{9}{25}$

Check: $4\sqrt{\frac{9}{25}} + \sqrt{1 + 16\left(\frac{9}{25}\right)} = 5$? Yes

$\left\{\frac{9}{25}\right\}$

22. $4x + 17 = (4 - \sqrt{x + 1})^2$
$4x + 17 = 16 - 8\sqrt{x + 1} + (x + 1)$
$3x = -8\sqrt{x + 1}$
$9x^2 = 64(x + 1)$
$9x^2 - 64x - 64 = 0$
$(9x + 8)(x - 8) = 0$
$x = \frac{-8}{9}$; $x = 8$

Check: $\sqrt{4\left(\frac{-8}{9}\right) + 17} = 4 - \sqrt{\frac{-8}{9} + 1}$? Yes
$\sqrt{4(8) + 17} = 4 - \sqrt{8 + 1}$? No

$\left\{\frac{-8}{9}\right\}$

24. $(y + 7)^{1/2} = 3 - (y + 4)^{1/2}$
$y + 7 = 9 - 6(y + 4)^{1/2} + (y + 4)$
$-6 = -6(y + 4)^{1/2}$
$(y + 4)^{1/2} = 1$
$y + 4 = 1$
$y = -3$

Check: $(-3 + 7)^{1/2} + (-3 + 4)^{1/2} = 3$? Yes
$\{-3\}$

26. $(z + 5)^{1/2} = 4 - (z - 3)^{1/2}$
$z + 5 = 16 - 8(z - 3)^{1/2} + (z - 3)$
$-8 = -8(z - 3)^{1/2}$
$(z - 3)^{1/2} = 1$
$z - 3 = 1$
$z = 4$

Check: $(4 - 3)^{1/2} + (4 + 5)^{1/2} = 4$? Yes
$\{4\}$

28. $t^2 = \dfrac{2v}{g}$

$g = \dfrac{2v}{t^2}$

30. $P^2 = \pi^2\left(\dfrac{\ell}{g}\right)$

$gP^2 = \pi^2\ell$

$g = \dfrac{\pi^2\ell}{P^2}$

32.

$$(q - 1)^2 = 4\left(\frac{r^2 - 1}{3}\right)$$

$$[q^2 - 2q + 1](3) = \left[4\left(\frac{r^2 - 1}{3}\right)\right](3)$$

$$3q^2 - 6q + 3 = 4r^2 - 4$$

$$3q^2 - 6q + 7 = 4r^2$$

$$r^2 = \frac{3q^2 - 6q + 7}{4}; \quad r = \frac{\pm\sqrt{3q2 - 6q + 7}}{2}$$

34. $\ell \geqslant 0$ and $g > 0$ or $\ell < 0$ and $g < 0$.

EXERCISE 6.7

2. Let $x^2 = u$, $x^4 = u^2$; then substitute for x^2 and x^4.

$u^2 - 13u + 36 = 0$

$(u - 9)(u - 4) = 0$

$u = 9$ or $u = 4$

Hence,

$x^2 = 9$ or $x^2 = 4$

$x = \pm 3$ or $x = \pm 2$

$\{-2, 2, 3, -3\}$

4. Let $z^2 = u$, $z^4 = u^2$; then substitute for z^2 and z^4.

$u^2 - 2u - 24 = 0$

$(u - 6)(u + 4) = 0$

$u = 6$ or $u = -4$

Hence,

$z^2 = 6$ or $z^2 = -4$

$z = \pm\sqrt{6}$ or $z = \pm 2i$

$\{-\sqrt{6}, \sqrt{6}, -2i, 2i\}$

6. Let $\sqrt{y} = u$, $y = u^2$; then substitute for y and $\sqrt{y}$.

$u^2 + 3u - 10 = 0$

$(u - 2)(u + 5) = 0$

$u = 2$ or $u = -5$

Hence,

$\sqrt{y}$ cannot equal -5;

$\sqrt{y} = 2$ or $y = 4$

Check: $4 + 3\sqrt{4} - 10 = 0$?

Yes $\{4\}$

8. Let $\sqrt{y^2 - 5} = u$, $y^2 - 5 = u^2$;
then substitute for $y^2 - 5$ and $\sqrt{y^2 - 5}$.

$$u^2 - 5u + 6 = 0$$
$$(u - 2)(u - 3) = 0$$
$$u = 2 \quad \text{or} \quad u = 3$$

Hence,

$$\sqrt{y^2 - 5} = 2 \quad \text{or} \quad \sqrt{y^2 - 5} = 3$$
$$y^2 - 5 = 4 \quad \text{or} \quad y^2 - 5 = 9$$
$$y^2 = 9 \quad \text{or} \quad y^2 = 14$$
$$y = \pm 3 \quad \text{or} \quad y = \pm\sqrt{14}$$

Check: $(\pm 3)^2 - 5 - 5\sqrt{(\pm 3)^2 - 5} + 6 = 0?$ Yes
$(\pm\sqrt{14})^2 - 5 - 5\sqrt{(\pm\sqrt{14})^2 - 5} + 6 = 0?$ Yes

$\{-3, 3, -\sqrt{14}, \sqrt{14}\}$

10. Let $z^{1/3} = u$, $z^{2/3} = u^2$
then substitute for $z^{1/3}$ and $z^{2/3}$.

$$u^2 - 2u - 35 = 0$$
$$(u - 7)(u + 5) = 0$$
$$u = 7 \quad \text{or} \quad u = -5$$

Hence,

$$z^{1/3} = 7 \quad \text{or} \quad z^{1/3} = -5$$
$$z = 343 \quad \text{or} \quad z = -125$$

Check: $(-125)^{2/3} - 2(-125)^{1/3} = 35?$
$25 - 2(-5) = 35?$ Yes
$(343)^{2/3} - 2(343)^{1/3} = 35?$
$49 - 2(7) = 35?$ Yes

$\{343, -125\}$

12. Let $y^{1/3} = u$, $y^{2/3} = u^2$;
then substitute for $y^{1/3}$ and $y^{2/3}$.

$$2u^2 + 5u = 3$$
$$2u^2 + 5u - 3 = 0$$
$$(2u - 1)(u + 3) = 0$$
$$u = \frac{1}{2} \quad \text{or} \quad u = -3$$

Hence,

$$y^{1/3} = \frac{1}{2} \quad \text{or} \quad y^{1/3} = -3$$
$$(y^{1/3})^3 = y = \left(\frac{1}{2}\right)^3 \quad \text{or} \quad (y^{1/3})^3 = y = (-3)^3$$
$$y = \frac{1}{8} \quad \text{or} \quad y = -27$$

$$\text{Check: } 2\left(\frac{1}{8}\right)^{2/3} + 5\left(\frac{1}{8}\right)^{1/3} = 3?$$
$$2\left(\frac{1}{4}\right) + 5\left(\frac{1}{2}\right) = 3?$$
$$\frac{1}{2} + \frac{5}{2} = 3? \quad \text{Yes}$$
$$2(-27)^{2/3} + 5(-27)^{1/3} = 3?$$
$$2(9) + 5(-3) = 3? \quad \text{Yes}$$

$\left\{-27, \frac{1}{8}\right\}$

14. Let $z^{1/2} = u$, $z = u^2$;
then substitute u^2 for z and u for $z^{1/2}$.

$$u^2 + u = 72$$
$$u^2 + u - 72 = 0$$
$$(u + 9)(u - 8) = 0$$
$$u = -9 \quad \text{or} \quad u = 8$$

Hence,

$z^{1/2} = 8$ or $z = 64$
$z^{1/2}$ cannot equal -9

Check: $64 + 64^{1/2} = 72$? Yes
$\{64\}$

16. Let $x^{1/4} = u$, $x^{1/2} = u^2$;
then substitute u^2 for $x^{1/2}$ and u for $x^{1/4}$.

$$8u^2 + 7u = 1$$
$$8u^2 + 7u - 1 = 0$$
$$(8u - 1)(u + 1) = 0$$
$$8u - 1 = 0 \quad \text{or} \quad u + 1 = 0$$
$$u = \frac{1}{8} \quad \text{or} \quad u = -1$$

Hence,

$x^{1/4} = \frac{1}{8}$ or $x = \left(\frac{1}{8}\right)^4 = \frac{1}{4096}$
$x^{1/4}$ cannot equal -1

Check: $8\left(\frac{1}{4096}\right)^{1/2} + 7\left(\frac{1}{4096}\right)^{1/4} = 1$? Yes

$\left\{\frac{1}{4096}\right\}$

18. Let $z^{-1} = u$, $z^{-2} = u^2$;

$$u^2 + 9u - 10 = 0$$
$$(u + 10)(u - 1) = 0$$
$$u = -10 \quad \text{or} \quad u = 1$$

Hence,

$z^{-1} = -10$ or $z^{-1} = 1$

$\frac{1}{z} = -10$ or $\frac{1}{z} = 1$

$z = \frac{-1}{10}$ $\quad z = 1$

Check: $\left(\frac{-1}{10}\right)^{-2} + 9\left(\frac{-1}{10}\right)^{-1} - 10 = 0$: Yes

$1^{-2} + 9(1)^{-2} - 10 = 0$? Yes

$\left\{\frac{-1}{10}, 1\right\}$

20. Let $(x - 2)^{1/4} = u$, $(x - 2)^{1/2} = u^2$; then substitute.

$u^2 - 11u + 18 = 0$
$(u - 9)(u - 2) = 0$
$u = 9$ or $u = 2$

Hence,

$(x - 2)^{1/4} = 9$ or $(x - 2)^{1/4} = 2$
$x - 2 = 9^4$ or $x - 2 = 2^4$
$x - 2 = 6541$ or $x - 2 = 16$
$x = 6543$ or $x = 18$

Check: $(6543 - 2)^{1/2} - 11(6543 - 2)^{1/4} + 18 = 0$? Yes
$(18 - 2)^{1/2} - 11(18 - 2)^{1/4} + 18 = 0$? Yes

$\{6543, 18\}$

22. a. $\sqrt{x} = 6 - x$
$x = (6 - x)^2$
$x = 36 - 12x + x^2$
$0 = x^2 - 13x + 36$
$0 = (x - 9)(x - 4)$
$x = 9$ or $x = 4$

Check:
$9 + \sqrt{9} - 6 = 0$? No
$4 + \sqrt{4} - 6 = 0$? Yes
$\{4\}$

b. Let $\sqrt{x} = u$, $x = u^2$;
$u^2 + u - 6 = 0$
$(u + 3)(u - 2) = 0$
$u = -3$ or $u = 2$
Hence,

$\sqrt{x} = 2$ or $x = 4$;
$\sqrt{x}$ cannot equal -3

Check: See part a.
$\{4\}$

24. a. $2\sqrt{y} = 15 - y$
$4y = 225 - 30y + y^2$
$0 = y^2 - 34y + 225$
$0 = (y - 25)(y - 9)$
$y = 25$ or $y = 9$

Check:
$25 + 2\sqrt{25} = 15$? No
$9 + 2\sqrt{9} = 15$? Yes
$\{9\}$

b. Let $\sqrt{y} = u$, $y = u^2$;
$u^2 + 2u = 15$
$u^2 + 2u - 15 = 0$
$(u + 5)(u - 3) = 0$
$u = -5$ or $u = 3$
Hence,

$\sqrt{y} = 3$ or $y = 9$;
$\sqrt{y}$ cannot equal -5

Check: See part a.
$\{9\}$

2. The critical numbers are -2 and -5 because $(x + 2)(x + 5) = 0$ when x has either of these values. The critical numbers are graphed with open dots because these numbers do not satisfy the given inequality.

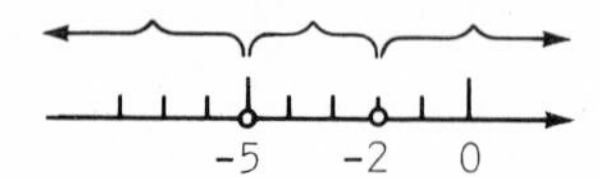

A number from each of the three intervals shown on the number line, say -6, -3, and 0, is substituted into the given inequality.

$$(-6 + 2)(-6 + 5) \overset{?}{<} 0; \quad (-3 + 2)(-3 + 5) \overset{?}{<} 0;$$
no yes

$$(0 + 2)(0 + 5) \overset{?}{<} 0$$
no

Hence, the solution set is: $\{x \mid -5 < x < -2\}$ or $(-5,-2)$.

The graph is: -5 -2 0

4. The critical numbers are 0 and -3 because $x(x + 3) = 0$ when x has either of these values. The critical numbers are graphed with closed dots because these numbers satisfy the given inequality.

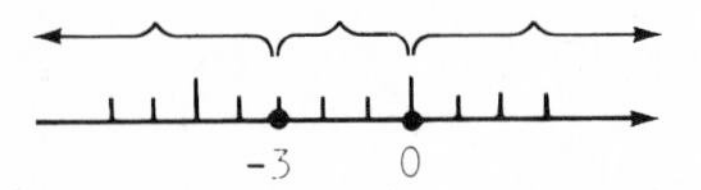

A number from each of three intervals shown on the number line, say -4, -1, and 1, is substituted into the given inequality.

$$-4(-4 + 3) \overset{?}{\geqslant} 0; \quad -1(-1 + 3) \overset{?}{\geqslant} 0; \quad 1(1 + 3) \overset{?}{\geqslant} 0$$
yes no yes

Hence, the solution set is: $\{x \mid x \leqslant -3\} \cup \{x \mid x \geqslant 0\}$ or $(-\infty,-3] \cup [0,+\infty)$.

The graph is: -3 0

6. The critical numbers are -1 and 6.

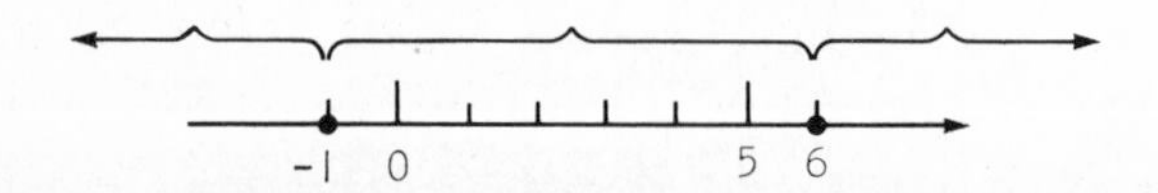

A number from each of the three intervals shown on the number line, say -2, 0, and 7, is substituted into the given inequality.

$$(-2)^2 - 5(-2) - 6 \overset{?}{\geqslant} 0; \quad 0^2 - 5(0) - 6 \overset{?}{\geqslant} 0;$$
yes no

$$7^2 - 5(7) - 6 \overset{?}{\geqslant} 0$$
yes

Hence, the solution set is: $\{x|x \leqslant -1\} \cup \{x|x \geqslant 6\}$ or $(-\infty,-1] \cup [6,+\infty)$.

The graph is:

8. Since every real number squared is nonnegative, $4x^2 + 1$ is greater than 0 for every real number value of x. Hence, the solution set is: $\emptyset$.
There are no points on the graph.

10. Since x^2 is greater than or equal to zero for every real number value of x, $x^2 + 1 \geqslant 0$ for every real number value of x. Hence, the solutions set is: $\{x|x$ is a real number$\}$ or $(-\infty,+\infty)$.

The graph is:

-5 0 5

12. $\dfrac{3}{x - 6} - 8 > 0; \quad \dfrac{3 - 8(x - 6)}{x - 6} > 0; \quad \dfrac{-8x + 51}{x - 6} > 0$

The critical numbers are $\dfrac{51}{8}$ and 6 because $\dfrac{-8x + 51}{x - 6} = 0$ when $x = \dfrac{51}{8}$ and because $\dfrac{-8x + 51}{x - 6}$ is undefined when $x = 6$. We graph the critical numbers with open dots because neither of them satisfies the given inequality.

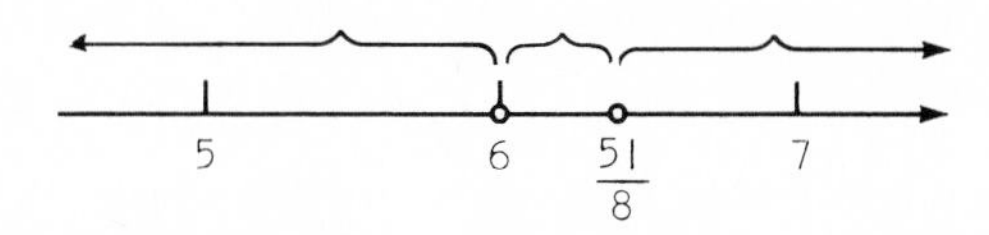

A number from each of the three intervals shown on the number line, say, 5, $\dfrac{50}{8} = \dfrac{25}{4}$, and 7, is substituted into the given inequality.

$$\frac{3}{5 - 6} \overset{?}{>} 8; \quad \frac{3}{\frac{25}{4} - \frac{24}{4}} \overset{?}{>} 8; \quad \frac{3}{7 - 6} \overset{?}{>} 8$$
no yes no

Hence, the solution set is: $\left\{ x \mid 6 < x < \frac{51}{8} \right\}$ or $\left(6, \frac{51}{8}\right)$.

The graph is:

14. $\frac{x + 2}{x - 2} - 6 \geqslant 0$; $\frac{x + 2 - 6(x - 2)}{x - 2} \geqslant 0$; $\frac{-5x + 14}{x - 2} \geqslant 0$

The critical numbers are $\frac{14}{5}$ and 2 because $\frac{-5x + 14}{x - 2} = 0$ when $x = \frac{14}{5}$ and because $\frac{-5x + 14}{x - 2}$ is undefined when $x = 2$. $\frac{14}{5}$ is graphed with a closed dot because it satisfies the given inequality, and 2 is graphed with an open dot because it doesn't.

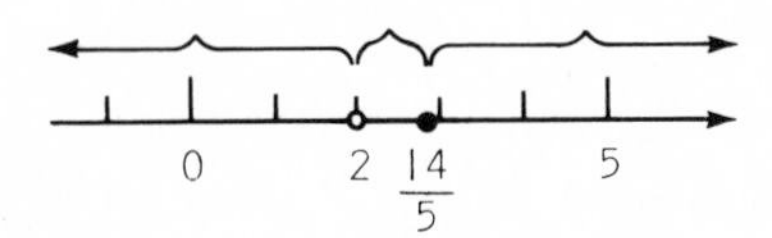

A number from each of the three intervals shown on the number line, say 0, $\frac{12}{5}$, and 3, is substituted into the given inequality.

$$\frac{0 + 2}{0 - 2} \overset{?}{\geqslant} 6; \quad \frac{\frac{12}{5} + \frac{10}{5}}{\frac{12}{5} - \frac{10}{5}} \overset{?}{\geqslant} 6; \quad \frac{3 + 2}{3 - 2} \overset{?}{\geqslant} 6$$

no yes no

Hence, the solution set is: $\left\{x \mid 2 < x \leqslant \frac{14}{5}\right\}$ or $\left(2, \frac{14}{5}\right]$.

The graph is:

16. $\frac{x}{x - 1} - 5 \geqslant 0$; $\frac{x - 5(x - 1)}{x - 1} \geqslant 0$; $\frac{-4x + 5}{x - 1} \geqslant 0$

The critical numbers are $\frac{5}{4}$ and 1 because $\frac{-4x + 5}{x - 1} = 0$ when $x = \frac{5}{4}$ and because $\frac{-4x + 5}{x - 1}$ is undefined when $x = 1$. $\frac{5}{4}$ is graphed with a closed dot because it satisfies the given inequality, and 1 is graphed with an open dot because it does not.

A number from each of the three intervals shown on the number line, say 0, $\frac{9}{8}$, and 2, is substituted into the given inequality.

$$\frac{0}{0-1} \overset{?}{\geqslant} 5; \text{ no} \qquad \frac{\frac{9}{8}}{\frac{9}{8}-1} = 9 \overset{?}{\geqslant} 5; \text{ yes} \qquad \frac{2}{2-1} \overset{?}{\geqslant} 5 \text{ no}$$

Hence, the solution set is: $\left\{x \middle| 1 < x \leqslant \frac{5}{4}\right\}$ or $\left(1, \frac{5}{4}\right]$.

The graph is:

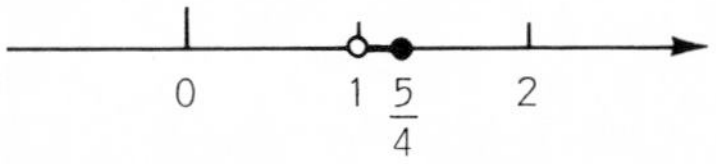

0 1 $\frac{5}{4}$ 2

18. $\frac{x+1}{x-1} - 1 < 0$; $\frac{x+1-1(x-1)}{x-1} < 0$; $\frac{2}{x-1} < 0$.

Because $\frac{2}{x-1}$ is undefined when $x = 1$, 1 is a critical number. No value of x can make $\frac{2}{x-1}$ equal zero; so 1 is the only critical number. 1 is graphed with an open dot because it does not satisfy the inequality.

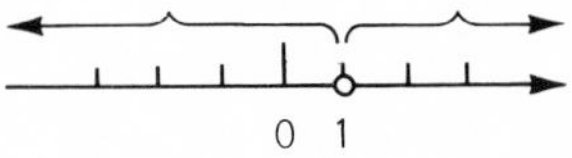

0 1

A number from each of the two intervals shown on the number line, say 0 and 2, is substituted into the given inequality.

$$\frac{0+1}{0-1} \overset{?}{<} 1; \text{ yes} \qquad \frac{2+1}{2-1} \overset{?}{<} 1 \text{ no}$$

Hence, the solution set is: $\{x | x < 1\}$ or $(-\infty, 1)$.

The graph is:

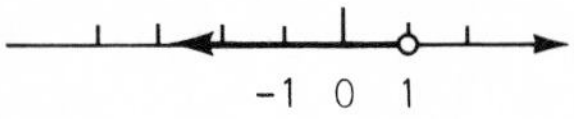

-1 0 1

20. $\frac{3}{x-1} - \frac{1}{x} \geqslant 0$; $\frac{3x-(x-1)}{x(x-1)} \geqslant 0$; $\frac{2x+1}{x(x-1)} \geqslant 0$

The critical numbers are $\frac{-1}{2}$, 0, and 1 because $\frac{2x+1}{x(x-1)} = 0$ when $x = \frac{-1}{2}$ and because $\frac{2x+1}{x(x-1)}$ is undefined when x equals either 0 or 1. $\frac{-1}{2}$ is graphed with a closed dot because it satisfies the given inequality, and 0 and 1 are graphed with open dots because they don't.

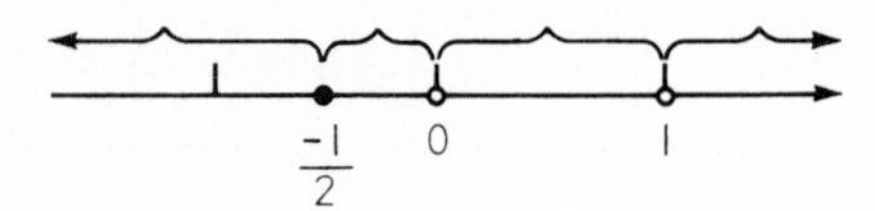

A number from each of the four intervals shown on the number line is substituted into the given inequality to determine the solution set: $\left\{x \middle| \frac{-1}{2} \leqslant x < 0\right\} \cup \{x| > 1\}$ or $\left[-\frac{1}{2}, 0\right) \cup (1, +\infty)$

The graph is:

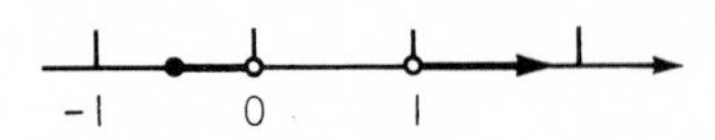

22. $\frac{3}{4x + 1} - \frac{2}{x - 5} > 0$; $\frac{3(x - 5) - 2(4x + 1)}{(4x + 1)(x - 5)} > 0$;

$\frac{-5x - 17}{(4x + 1)(x - 5)} > 0$

The critical numbers are $\frac{-17}{5}$, $\frac{-1}{4}$, and 5 because $\frac{-5x - 17}{(4x + 1)(x - 5)} = 0$ when $x = \frac{-17}{5}$ and because $\frac{-5x - 17}{(4x + 1)(x - 5)}$ is undefined when x equals either $\frac{-1}{4}$ or 5. All of these critical numbers are graphed with open dots.

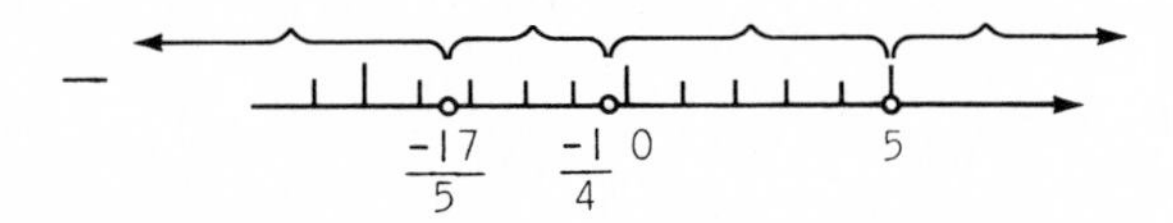

A number from each of the four intervals shown on the number line is substituted into the given inequality to determine the solution set: $\left\{x \middle| x < \frac{-17}{5}\right\} \cup \left\{x \middle| -\frac{1}{4} < x < 5\right\}$ or $\left(-\infty, \frac{-17}{5}\right) \cup \left(-\frac{1}{4}, 5\right)$.

The graph is:

$\frac{-17}{5}$ $\frac{-1}{4}$ 0 5

24. The given inequality is equivalent to

$$-2 < y^2 - 3 \quad \text{and} \quad y^2 - 3 < 13$$

which can be written equivalently as

$$0 < y^2 - 1 \quad \text{and} \quad y^2 - 16 < 0.$$

The critical numbers are those values of y that are solutions of $y^2 - 1 = 0$ or $y^2 - 16 = 0$. These are 1, -1, 4, -4, which are graphed with open dots since they do not satisfy the given inequality.

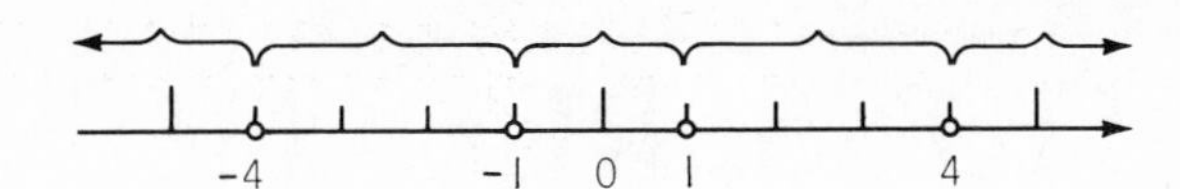

A number from each of the five intervals shown on the number line is substituted into the given inequality to determine the solution set: $\{y \mid -4 < y < -1\} \cup \{y \mid 1 < y < 4\}$ or $(-4,-1) \cup (1,4)$.

The graph is:

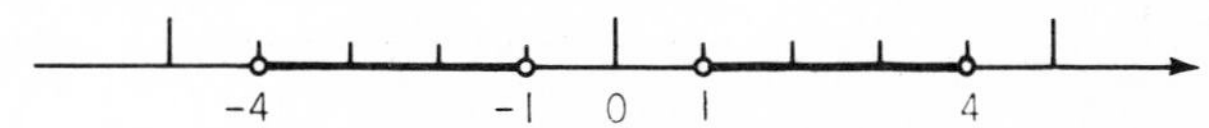

7

FUNCTIONS, RELATIONS, AND THEIR GRAPHS: PART I

EXERCISE 7.1

2. a. $y = 6 - 2(0) = 6;$ $(0,6)$

   b. $0 = 6 - 2x;$ $(3,0)$

   c. $y = 6 - 2(-1)$
   $= 6 + 2 = 8;$
   $(-1,8)$

4. Solve for y: $y = \dfrac{5 - x}{2}$

   a. $y = \dfrac{5 - (0)}{2} = \dfrac{5}{2};$ $\left(0, \dfrac{5}{2}\right)$

   b. $y = \dfrac{5 - (5)}{2} = 0;$ $(5,0)$

   c. $y = \dfrac{5 - (-3)}{2} = 4;$ $(-3,4)$

6. For $x = -2$,
   $y = 2(-2) + 6 = 2$

   for $x = 0$,
   $y = 2(0) + 6 = 6$

   for $x = 2$,
   $y = 2(2) + 6 = 10$

   $\{(-2,2),(0,6),(2,10)\}$

8. For $x = 5$,
   $y = 6 - 3(5) = -9$

   for $x = 10$,
   $y = 6 - 3(10) = -24$

   for $x = 15$,
   $y = 6 - 3(15) = -39$

   $\{(5,-9),(10,-24),(15,-39)\}$

10. Solve for y: $y = \dfrac{-2x + 12}{3}$

    For $x = \dfrac{-1}{4}$, $y = \dfrac{-2\left(\frac{-1}{4}\right) + 12}{3} = \dfrac{\frac{1}{2} + \frac{24}{2}}{3} = \dfrac{25}{6}$

    for $x = 0$, $y = \dfrac{-2(0) + 12}{3} = 4$

for $x = \frac{1}{4}$, $y = \dfrac{-2\left(\frac{1}{4}\right) + 12}{3} = \dfrac{\frac{-1}{2} + \frac{24}{2}}{3} = \frac{23}{6}$

$\left\{\left(\frac{-1}{4}, \frac{25}{6}\right), (0,4), \left(\frac{1}{4}, \frac{23}{6}\right)\right\}$

12. If $(-6,1)$ is a solution of $kx - 2y = 6$, then

$$k(-6) - 2(1) = 6,$$
$$-6k = 8, \quad k = \frac{-4}{3}.$$

14.
$$-xy = 6 - 3x$$
$$y = \frac{6 - 3x}{-x} = \frac{(-1)(6 - 3x)}{(-1)(-x)} = \frac{3x - 6}{x}$$

16.
$$x^2y - xy = -5$$
$$(x^2 - x)y = -5$$
$$y = \frac{-5}{x^2 - x} \quad \text{or} \quad y = \frac{5}{x - x^2}$$

18.
$$x^2y - xy - 5y = -3$$
$$(x^2 - x - 5)y = -3$$
$$y = \frac{-3}{x^2 - x - 5}$$

20.
$$3(y^2 + 1) = x$$
$$3y^2 + 3 = x$$
$$3y^2 = x - 3$$
$$y^2 = \frac{x - 3}{3}$$
$$y = \pm\sqrt{\frac{x - 3}{3}} \quad \text{or} \quad \frac{\pm\sqrt{3(x - 3)}}{3}$$

22.
$$x^2y + 2y = 3x^2$$
$$(x^2 + 2)y = 3x^2$$
$$y = \frac{3x^2}{x^2 + 2}$$

24. $y = (-1)^2 - 5(-1) + 7 = 13$

The solution is $(-1,13)$.

26.
$$1 = x^2 - 5x + 7$$
$$0 = x^2 - 5x + 6$$
$$0 = (x - 3)(x - 2)$$
$$x = 3 \quad \text{or} \quad x = 2$$

The solutions are $(3,1)$ and $(2,1)$.

28. $y = |-2| + 5 = 2 + 5 = 7$

The solution is $(-2,7)$.

30.
$$y = |2(-3) + 1| - |-3|$$
$$= |-5| - 3 = 5 - 3$$
$$= 2$$

The solution is $(-3,2)$.

32. $y = 2^{-3} = \frac{1}{2^3} = \frac{1}{8}$

The solution is $\left(-3, \frac{1}{8}\right)$.

34. $y = 2^{-(-3)} = 2^3 = 8$

The solution is $(-3,8)$.

EXERCISE 7.2

2.

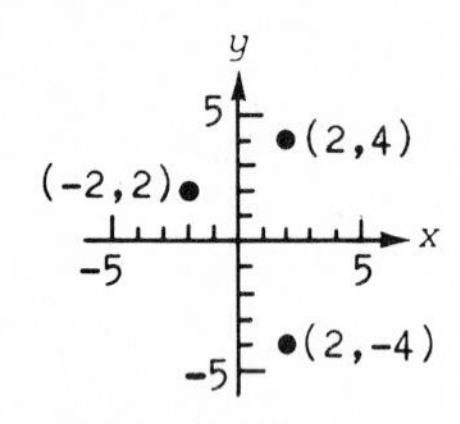

4.

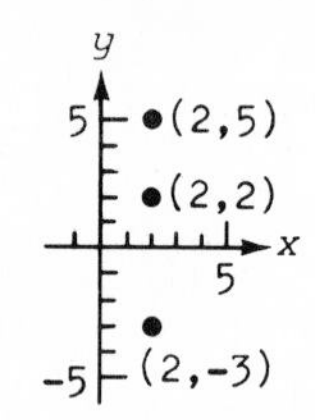

6.

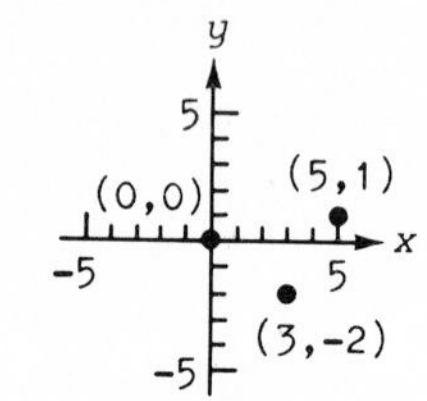

8.

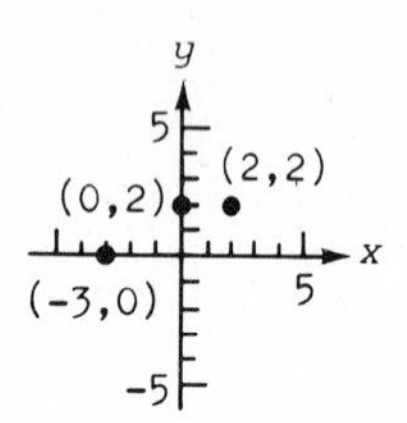

10. $x = 0, \quad y = (0) - 3 = -3$
$x = 2, \quad y = (2) - 3 = -1$
$x = 4, \quad y = (4) - 3 = 1$

Graph the ordered pairs: $(0,-3)$, $(2,-1)$, $(4,1)$.

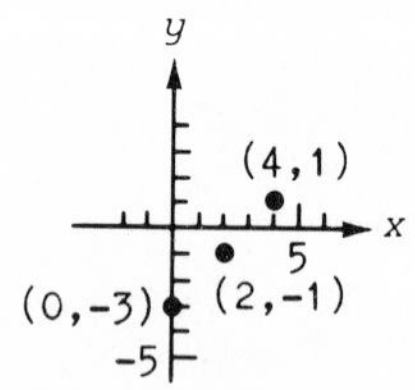

12. $x = -2, \quad y = 3(-2) - 2 = -8$
$x = -1, \quad y = 3(-1) - 2 = -5$
$x = 0, \quad y = 3(0) - 2 = -2$
$x = 1, \quad y = 3(1) - 2 = 1$
$x = 2, \quad y = 3(2) - 2 = 4$

Graph the ordered pairs: $(-2,-8)$, $(-1,-5)$, $(0,-2)$, $(1,1)$, $(2,4)$.

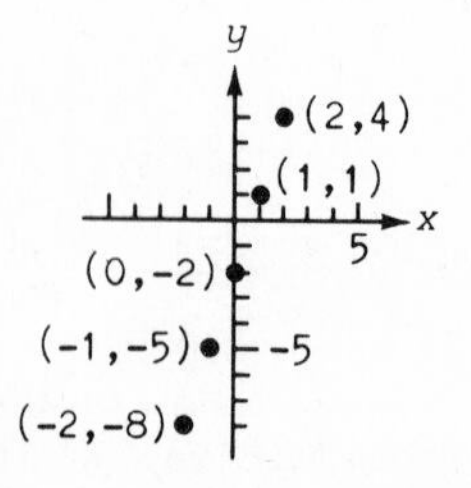

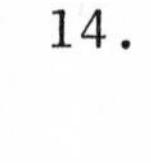

14. $x = -2, \quad y = 3 - 2(-2) = 7$
$x = -1, \quad y = 3 - 2(-1) = 5$
$x = 0, \quad y = 3 - 2(0) = 3$
$x = 1, \quad y = 3 - 2(1) = 1$
$x = 2, \quad y = 3 - 2(2) = -1$

Graph the ordered pairs: $(-2,7)$, $(-1,5)$, $(0,3)$, $(1,1)$, $(2,-1)$.

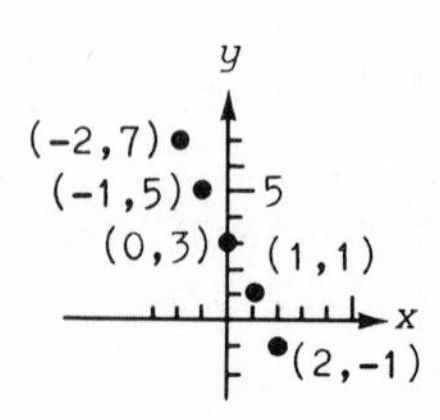

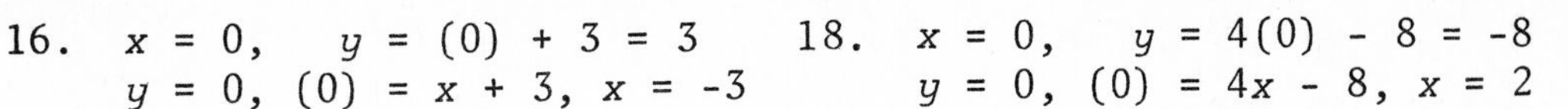

16. $x = 0, \quad y = (0) + 3 = 3$
$y = 0, \ (0) = x + 3, \ x = -3$

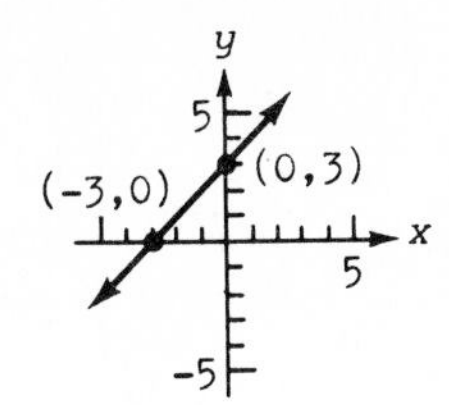

18. $x = 0, \quad y = 4(0) - 8 = -8$
$y = 0, \ (0) = 4x - 8, \ x = 2$

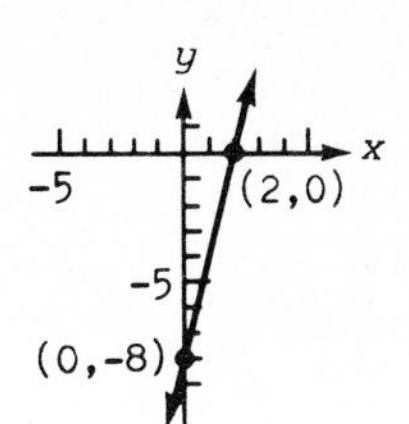

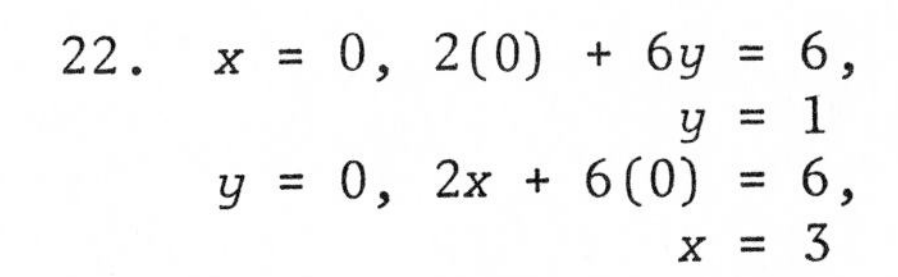

20. $x = 0, \ 2(0) - y = 6,$
$y = -6$
$y = 0, \ 2x - 0 = 6,$
$x = 3$

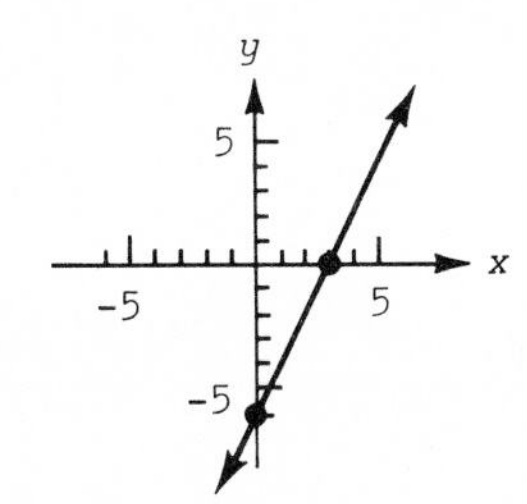

22. $x = 0, \ 2(0) + 6y = 6,$
$y = 1$
$y = 0, \ 2x + 6(0) = 6,$
$x = 3$

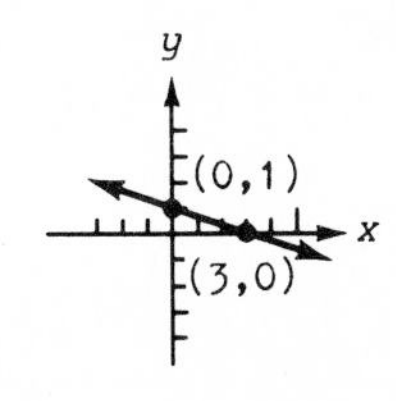

24. $x + 3y = 0$
$x = 0, \quad (0) + 3y = 0$
$y = 0$

Arbitrarily, let $y = -1$; then

$x + 3(-1) = 0$
$x = 3.$

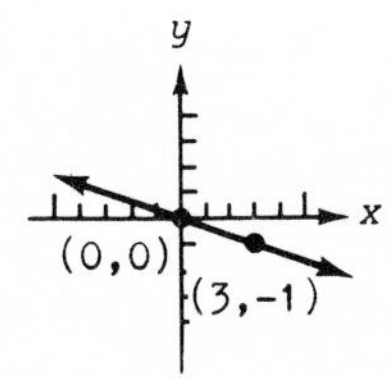

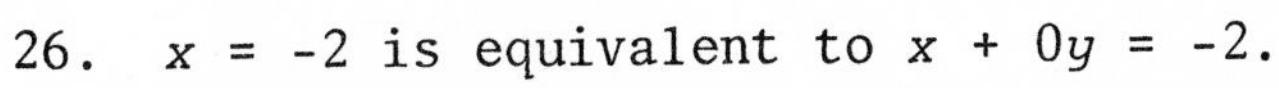

Graph (0,0) and (3,-1), and any other solution of the equation, and draw the line.

26. $x = -2$ is equivalent to $x + 0y = -2$.

For all values of y, $x = -2$. Graph any two ordered pairs of the form $(-2,y)$ and draw the line containing the two points.

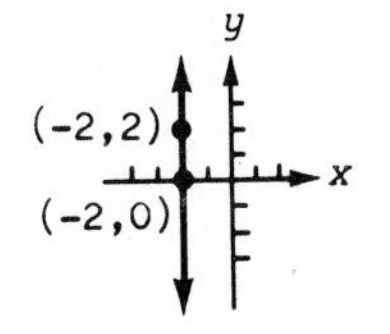

28. $y = 5$ is equivalent to $0x + y = 5$.

For all values of x, $y = 5$. Graph ny two ordered pairs of the form $(x,5)$ and draw the line.

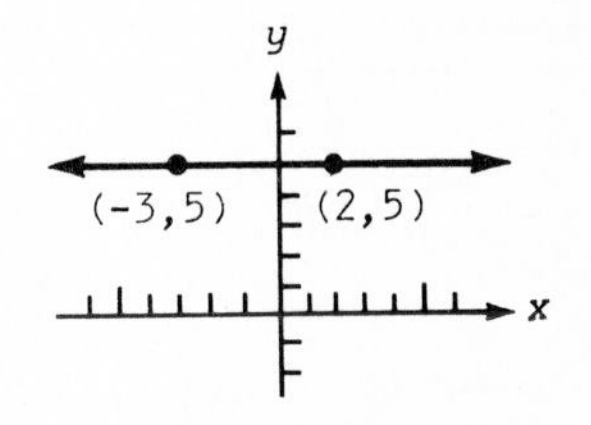

30. $y = 0$; $x = 2(0) = 0$

Arbitrarily, let $y = 3$; hence,

$$x = 2(3) = 6.$$

Graph (0,0) and (6,3), or any other solutions of the equation, and draw the line containing the two points.

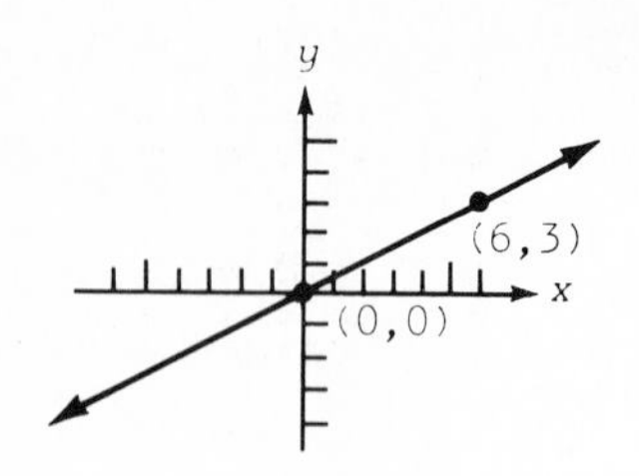

32. $x = 0$; $4(0) + y = 0$, $y = 0$

Arbitrarily, let $x = -1$; hence,

$$4(-1) + y = 0; \quad y = 4.$$

Graph (0,0) and (-1,4), or any other solutions of the equation, and draw the line containing the two points.

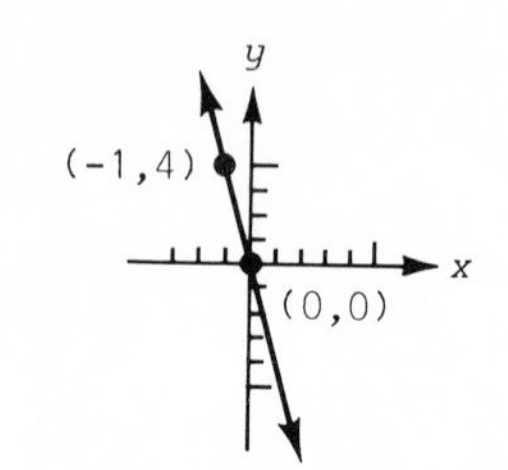

34. An open circle at (0,0) indicates that the point is not included. The dashed portion is not part of the graph.

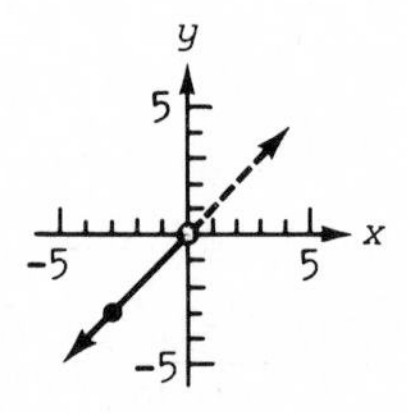

36. The dashed portion is not part of the graph.

38. The dashed portion is not part of the graph.

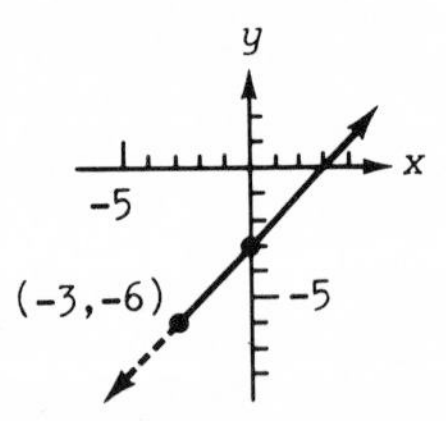

40. The dashed portion is not part of the graph.

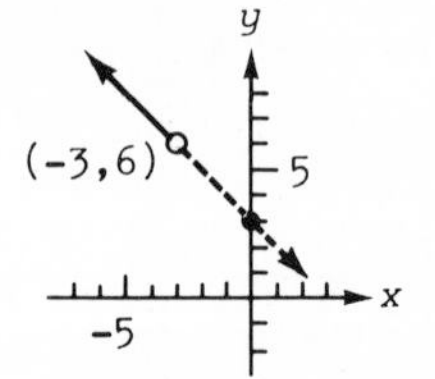

42. $y = 2|x|$ is equivalent to
$y = 2x$, if $x \geqslant 0$,
and
$y = 2(-x)$ or $y = -2x$, if $x < 0$.
The dashed portion is not part of the graph.

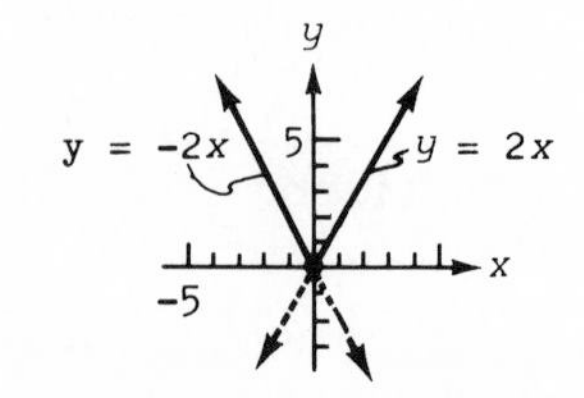

44. $y = |x - 2|$ is equivalent to
$y = x - 2$ if $x - 2 \geqslant 0$, or $x \geqslant 2$;
and
$y = -(x - 2) = -x + 2$ if $x - 2 < 0$, or $x < 2$.
The dashed portion is not part of the graph.

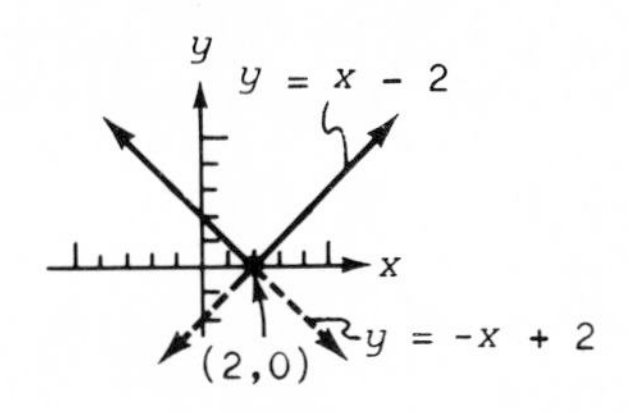

46. $y = |x| - 2$ is equivalent to
$y = x - 2$ if $x \geqslant 0$
and
$y = -x - 2$ if $x < 0$.
The dashed portion is not part of the graph.

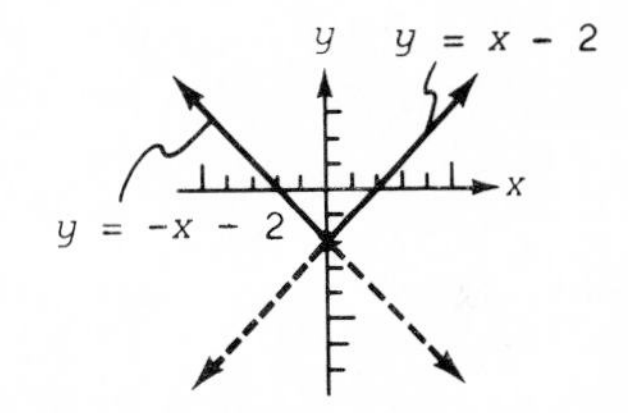

EXERCISE 7.3

2. a. Domain, the set of first components: $\{-1,0,1,2\}$
Range, the set of second components: $\{-1,0,1,2\}$

b. It is a function because no two ordered pairs have the same first component.

4. a. Domain: $\{4\}$
Range: $\{1,2,3,4\}$

b. Not a function since two or more pairs have the same first component.

6. a. Domain: $\{-5,2,3,6\}$
Range: $\{-5,2,3,6\}$

b. It is a function.

8. a. Domain: $\{2,3,4\}$
Range: $\{-1,1,2\}$

b. Not a function since $(3,-1)$ and $(3,1)$ have the same first component.

10.

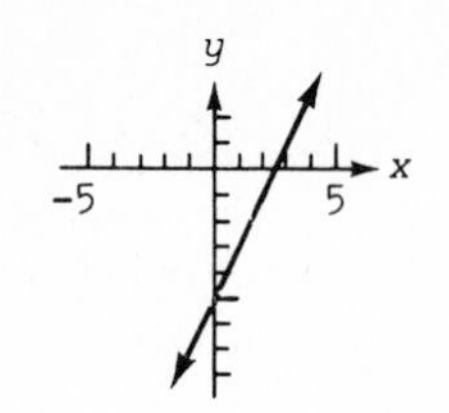

12.

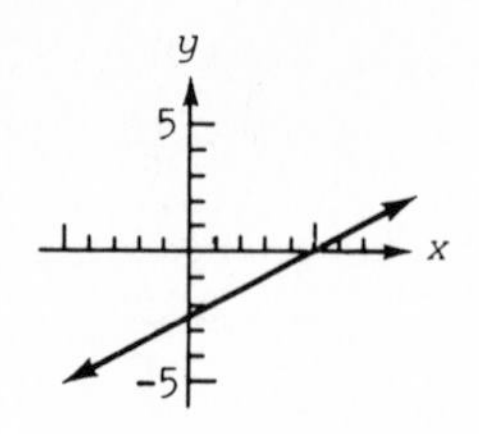

14.

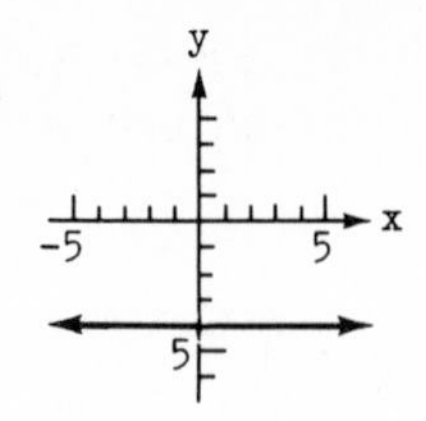

16. $g(3)$
$= 2(3)^2 + 3(3) - 1$
$= 26$

18. $f(2) = 3(2) - 1 = 5$
$f(0) = 3(0) - 1 = -1$
Hence,
$f(2) - f(0) = 5 - (-1) = 6.$

20. $f(0) = (0)^2 + 3(0) - 2 = -2$
$f(-2) = (-2)^2 + 3(-2) - 2 = -4$
Hence,
$f(0) - f(-2) = -2 - (-4) = 2.$

22. a. $f(x + h) = 2(x + h) + 5 = 2x + 2h + 5$

b. $f(x + h) - f(x) = (2x + 2h + 5) - (2x + 5) = 2h$

c. $\dfrac{f(x + h) - f(x)}{h} = \dfrac{2h}{h} = 2$

24. a. $f(x + h) = (x + h)^2 + 2(x + h) = x^2 + 2hx + h^2 + 2x + 2h$

b. $f(x + h) - f(x) = (x^2 + 2hx + h^2 + 2x + 2h) - (x^2 + 2x)$
$= 2hx + h^2 + 2h$

c. $\dfrac{f(x + h) - f(x)}{h} = \dfrac{2hx + h^2 + 2h}{h} = 2x + h + 2$

26. a. $f(x + h) = (x + h)^3 + 3(x + h)^2 - 1$
$= x^3 + 3x^2h + 3xh^2 + h^3 + 3x^2 + 6xh + 3h^2 - 1$

b. $f(x + h) - f(x)$
$= (x^3 + 3x^2h + 3xh^2 + h^3 + 3x^2 + 6xh + 3h^2 - 1)$
$- (x^3 + 3x^2 - 1)$
$= 3x^2h + 3xh^2 + h^3 + 6xh + 3h^2$

c. $\dfrac{f(x + h) - f(x)}{h} = \dfrac{3x^2h + 3xh^2 + h^3 + 6xh + 3h^2}{h}$
$= 3x^2 + 3xh + h^2 + 6x + 3h$

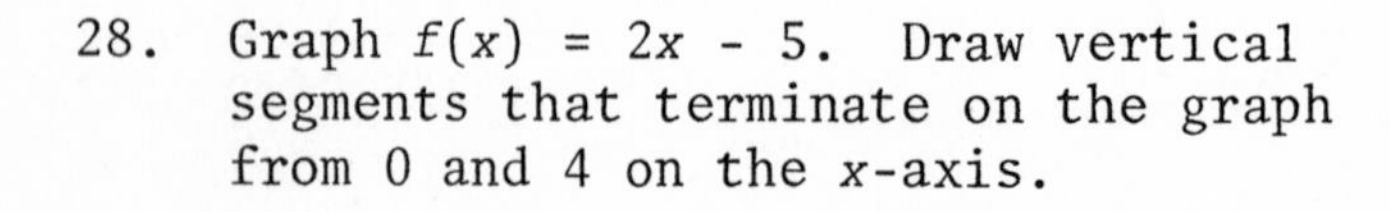

28. Graph $f(x) = 2x - 5$. Draw vertical segments that terminate on the graph from 0 and 4 on the x-axis.

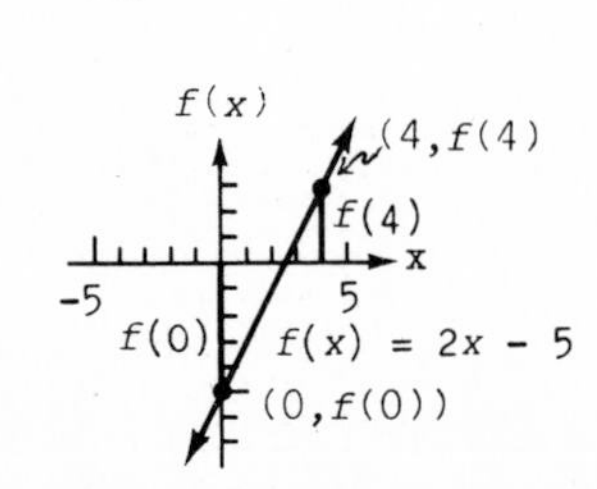

30. Graph $f(x) = 3x - 6$. Draw vertical segments that terminate on the graph from -3 and 6 on the x-axis.

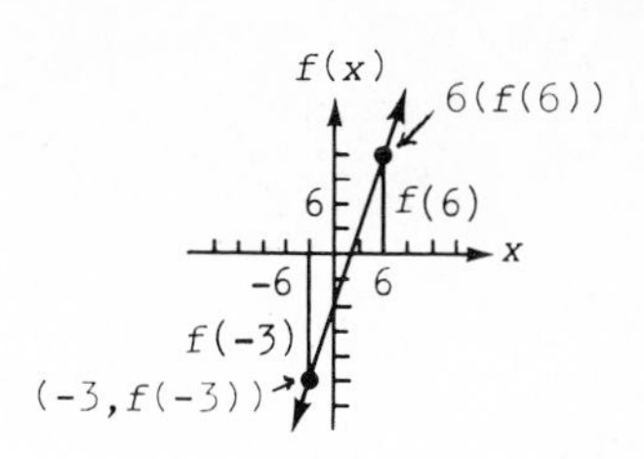

32. $\dfrac{BC}{AC} = \dfrac{f(x + h) - f(x)}{h}$

EXERCISE 7.4

2. $p_1 = (-1,1)$ and $p_2 = (5,9)$

$$m = \frac{9 - 1}{5 - (-1)} = \frac{8}{6} = \frac{4}{3}$$

$$d = \sqrt{[5 - (-1)]^2 + (9 - 1)^2}$$
$$= \sqrt{6^2 + 8^2} = \sqrt{100} = 10$$

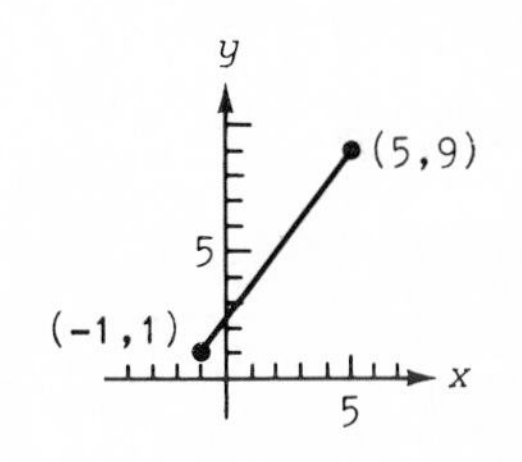

4. $p_1 = (-4,-3)$ and $p_2 = (1,9)$

$$m = \frac{9 - (-3)}{1 - (-4)} = \frac{12}{5}$$

$$d = \sqrt{[1 - (-4)]^2 + [9 - (-3)]^2}$$
$$\sqrt{5^2 + 12^2} = \sqrt{169} = 13$$

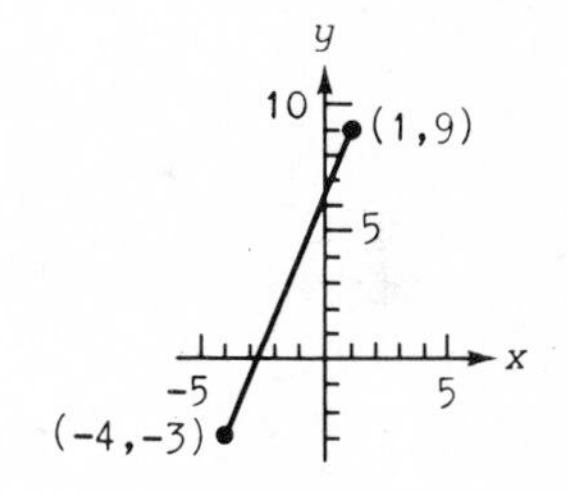

6. $p_1 = (-3,2)$ and $p_2 = (0,0)$

$$m = \frac{0 - 2}{0 - (-3)} = \frac{-2}{3}$$

$$d = \sqrt{[0 - (-3)]^2 + (0 - 2)^2}$$
$$= \sqrt{3^2 + (-2)^2} = \sqrt{13}$$

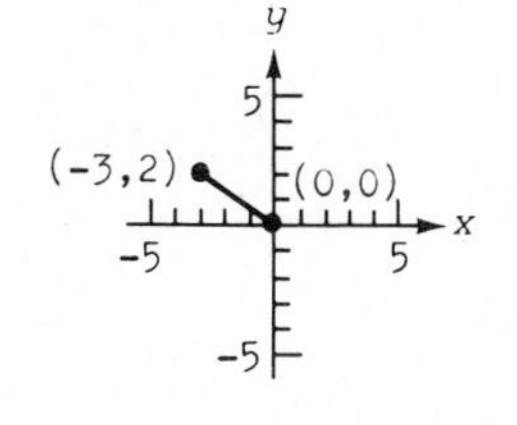

8. $p_1 = (2,-3)$ and $p_2 = (-2,-1)$

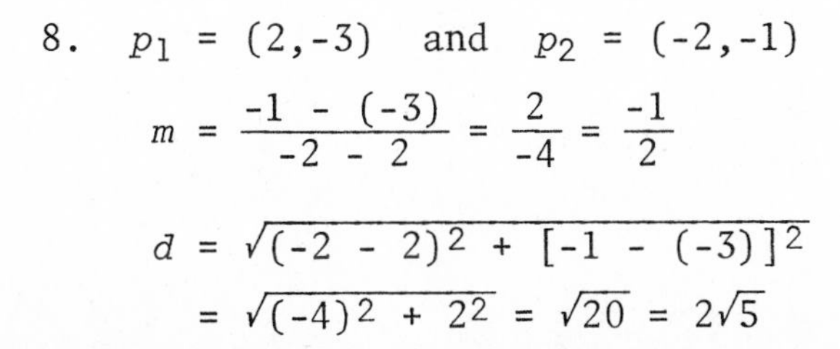

$$m = \frac{-1 - (-3)}{-2 - 2} = \frac{2}{-4} = \frac{-1}{2}$$

$$d = \sqrt{(-2 - 2)^2 + [-1 - (-3)]^2}$$
$$= \sqrt{(-4)^2 + 2^2} = \sqrt{20} = 2\sqrt{5}$$

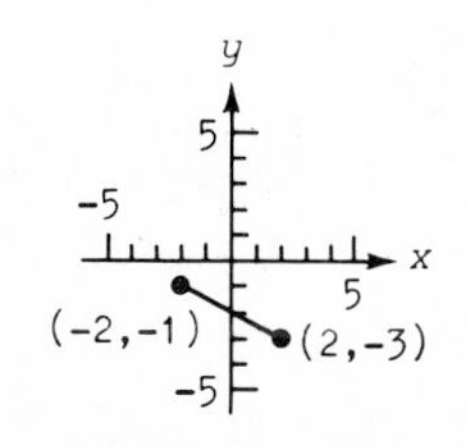

10. $p_1 = (2,0)$ and $p_2 = (-2,0)$

$$m = \frac{0 - 0}{-2 - 2} = 0$$

$$d = \sqrt{(-2 - 2)^2 + (0 - 0)^2} = \sqrt{(-4)^2} = 4$$

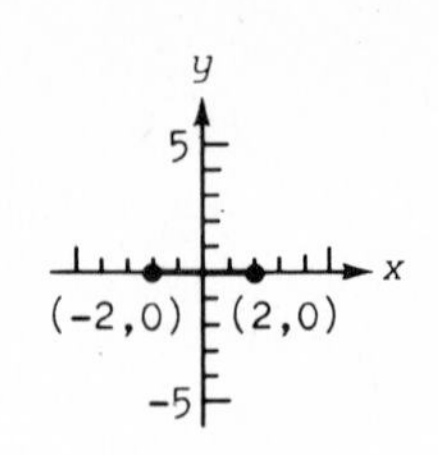

12. $p_1 = (-2,-5)$ and $p_2 = (-2,3)$

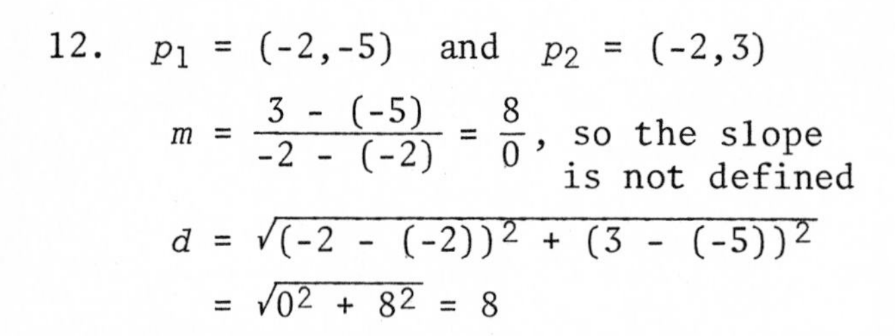

$$m = \frac{3 - (-5)}{-2 - (-2)} = \frac{8}{0},$$ so the slope is not defined

$$d = \sqrt{(-2 - (-2))^2 + (3 - (-5))^2}$$
$$= \sqrt{0^2 + 8^2} = 8$$

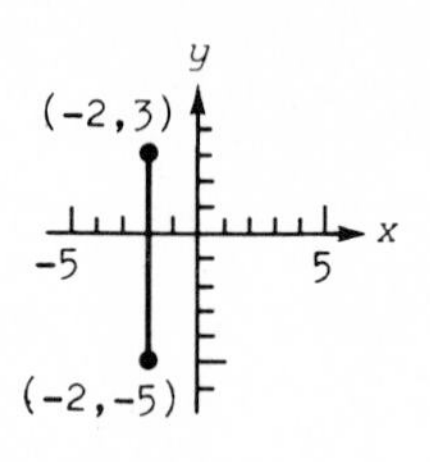

In Problems 14-16, let the points be A, B, and C, respectively, and P the perimeter.

14. $$AB = \sqrt{(3 - 10)^2 + (1 - 1)^2} = \sqrt{(-7)^2 + 0^2}$$
$$= \sqrt{49} = 7$$
$$AC = \sqrt{(5 - 10)^2 + (9 - 1)^2} = \sqrt{(-5)^2 + 8^2}$$
$$= \sqrt{89}$$
$$BC = \sqrt{(5 - 3)^2 + (9 - 1)^2} = \sqrt{2^2 + 8^2}$$
$$= \sqrt{68} = 2\sqrt{17}$$
$$P = 7 + \sqrt{89} + 2\sqrt{17}$$

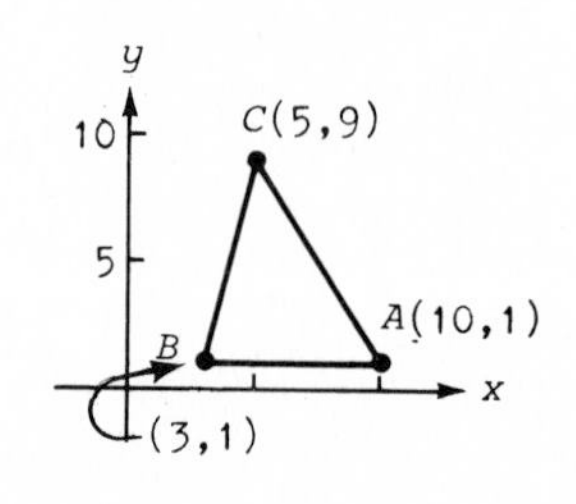

16.

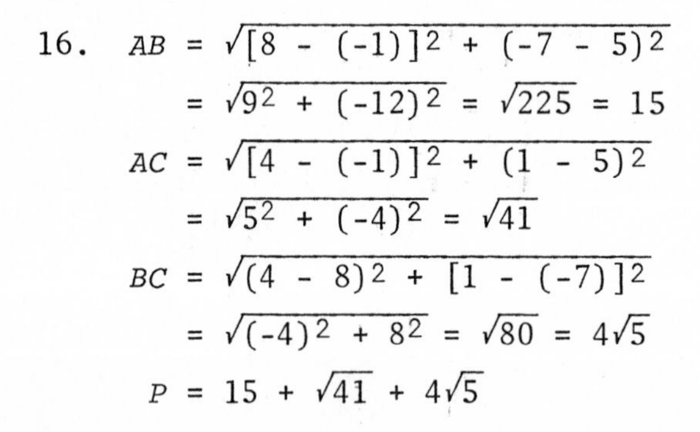

$$AB = \sqrt{[8 - (-1)]^2 + (-7 - 5)^2}$$
$$= \sqrt{9^2 + (-12)^2} = \sqrt{225} = 15$$
$$AC = \sqrt{[4 - (-1)]^2 + (1 - 5)^2}$$
$$= \sqrt{5^2 + (-4)^2} = \sqrt{41}$$
$$BC = \sqrt{(4 - 8)^2 + [1 - (-7)]^2}$$
$$= \sqrt{(-4)^2 + 8^2} = \sqrt{80} = 4\sqrt{5}$$
$$P = 15 + \sqrt{41} + 4\sqrt{5}$$

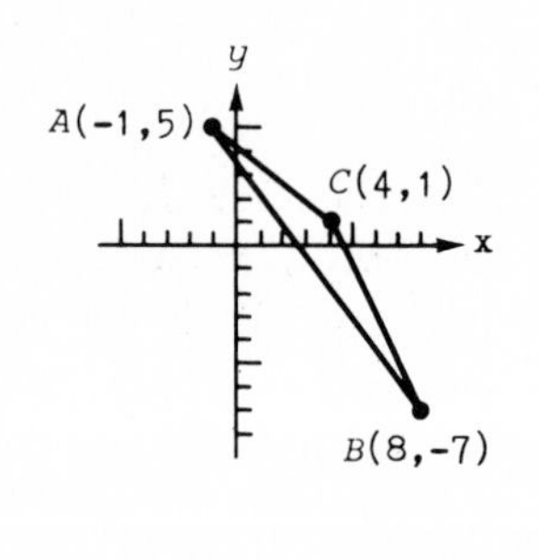

18. Let m_1 be the slope of the first segment, and m_2 the slope of the second segment. Then

$$m_1 = \frac{-2 - 2}{2 - (-4)} = \frac{-4}{6} = \frac{-2}{3}; \quad m_2 = \frac{4 - 0}{-3 - 3} = \frac{4}{-6} = \frac{-2}{3}.$$

Since $m_1 = m_2$, the segments are parallel.

20. Let m_1 be the slope of the segment joining (8,0) and (6,6), so $m_1 = \frac{6 - 0}{6 - 8} = \frac{6}{-2} = -3$; let m_2 be the slope of the segment with endpoints (-3,3) and (6,6), so

$$m_2 = \frac{6 - 3}{6 - (-3)} = \frac{3}{9} = \frac{1}{3}.$$

Since $m_1 m_2 = (-3)\left(\frac{1}{3}\right) = -1$ the line segments are perpendicular.

22.

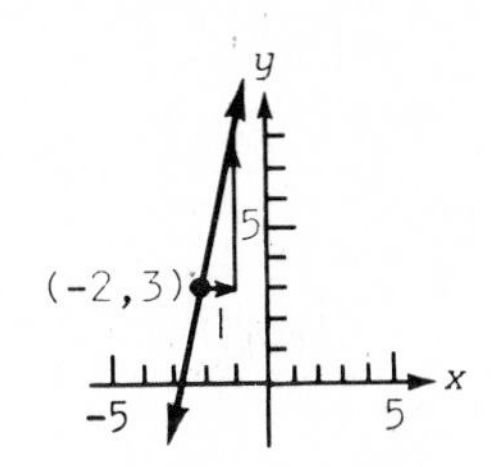

24.

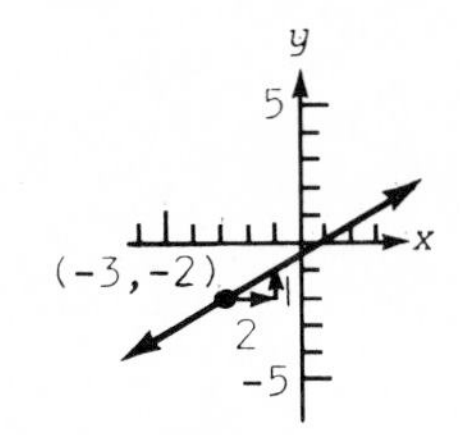

26.

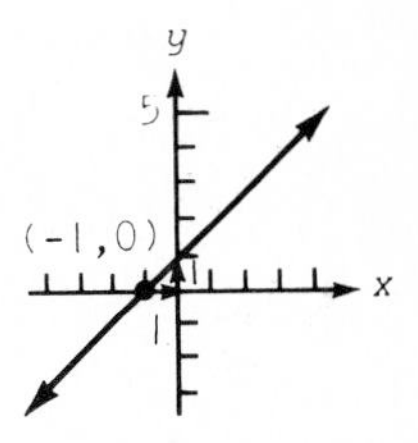

28.

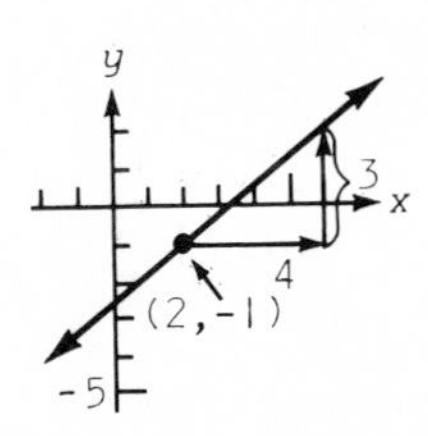

30.

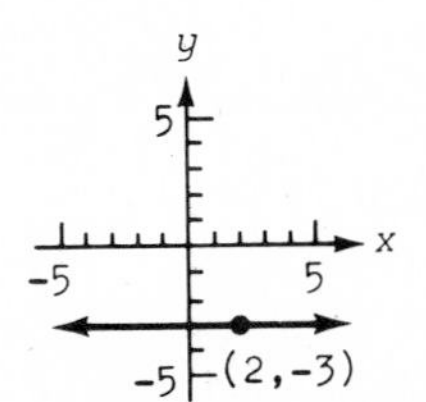

32.

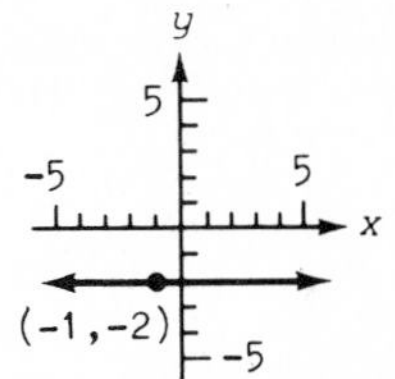

34. Let $A = (0,0)$, $B = (6,0)$, and $C = (3,3)$.

$$AB = \sqrt{(6 - 0)^2 + (0 - 0)^2} = \sqrt{36} = 6$$

$$AC = \sqrt{(3 - 0)^2 + (3 - 0)^2} = \sqrt{3^2 + 3^2} = \sqrt{18} = 3\sqrt{2}$$

$$BC = \sqrt{(3 - 6)^2 + (3 - 0)^2} = \sqrt{(-3)^2 + 3^2} = \sqrt{18} = 3\sqrt{2}$$

Since $(3\sqrt{2})^2 + (3\sqrt{2})^2 = 18 + 18 = 36$ and $6^2 = 36$, angle C is a right angle, and therefore, the triangle is a right triangle. $AC = BC$, so it is an isosceles triangle.

36. Let A, B, C, and D be the given points.

$$\text{slope of } AB = m_1 = \frac{-11 - 4}{7 - (-5)} = \frac{-15}{12} = \frac{-5}{4}$$

$$\text{slope of } CD = m_2 = \frac{40 - 25}{0 - 12} = \frac{15}{-12} = \frac{-5}{4}$$

$$\text{slope of } AD = m_3 = \frac{40 - 4}{0 - (-5)} = \frac{36}{5}$$

$$\text{slope of } BC = m_4 = \frac{25 - (-11)}{12 - 7} = \frac{36}{5}$$

Since $m_1 = m_2$ and $m_3 = m_4$, AB is parallel to CD and AD is parallel to BC so that $ABCD$ is a parallelogram.

38. If the segments are to be perpendicular, then $m_1 m_2 = -1$, or

$$(-4)\frac{k-4}{8} = -1; \quad \frac{k-4}{-2} = -1; \quad k - 4 = 2; \quad k = 6.$$

EXERCISE 7.5

Use the point-slope form, $y - y_1 = m(x - x_1)$, in Problems 2-12.

2. $y - (-5) = -3(x - 2)$
   $y + 5 = -3x + 6$
   $3x + y - 1 = 0$

4. $y - (-1) = 4[x - (-6)]$
   $y + 1 = 4x + 24$
   $-4x + y - 23 = 0$
   or
   $4x - y + 23 = 0$

6. $y - 0 = \frac{-1}{3}(x - 2)$
   $3y = -x + 2$
   $x + 3y - 2 = 0$

8. $y - (-1) = \frac{5}{3}(x - 2)$
   $3y + 3 = 5x - 10$
   $-5x + 3y + 13 = 0$
   or
   $5x - 3y - 13 = 0$

10. $y - (-6) = 0(x - 0)$
    $y + 6 = 0$

12. $y - 0 = 1(x - 0)$
    $y = x$
    $-x + y = 0$ or $x - y = 0$

In Problems 14-22, m = slope and b = y-intercept.

14. $y = -2x - 1$
    $m = -2$ and $b = -1$

16. $y = 3x - 7$
    $m = 3$ and $b = -7$

18. $y = \frac{2}{3}x$
    $m = \frac{2}{3}$ and $b = 0$

20. $y = \frac{-1}{2}x + \frac{5}{2}$
    $m = \frac{-1}{2}$ and $b = \frac{5}{2}$

22. $y = 3$
    $y = 0x + 3$
    $m = 0$ and $b = 3$

24. The line is parallel to the y-axis and has no y-intercept; the slope is not defined.

26. Solving $2y - 3x = 5$ explicitly for y gives $y = \frac{3}{2}x + \frac{5}{2}$; so the slope is $\frac{3}{2}$. Applying the point-slope form, $y - y_1 = m(x - x_1)$, using (0,5) and $m = \frac{3}{2}$,

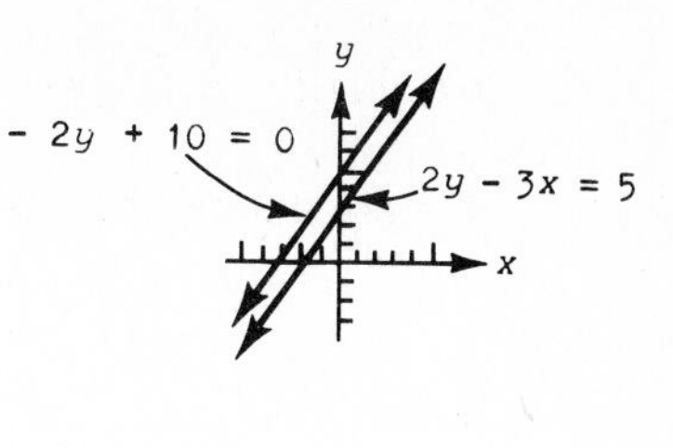

$$y - 5 = \frac{3}{2}(x - 0)$$
$$3x - 2y + 10 = 0.$$

28. Solve $2y - 3x = 5$ for y: $y = \frac{3}{2}x + \frac{5}{2}$. The slope of this line is $\frac{3}{2}$. The slope of a perpendicular is $\frac{-2}{3}$. Using the point-slope form:

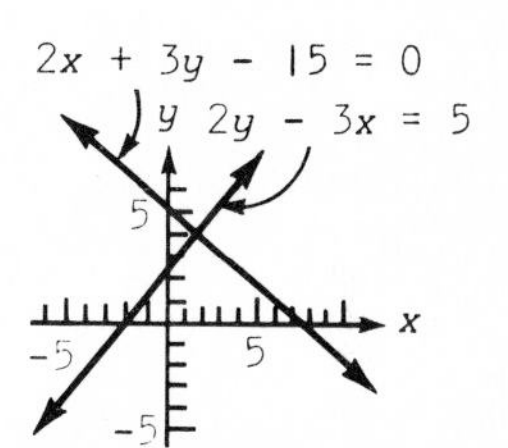

$$y - 5 = \frac{-2}{3}(x - 0)$$
$$2x + 3y - 15 = 0.$$

30. Let the first point be (x_1,y_1) and the second be (x_2,y_2).

a. $y - 1 = \left(\frac{3 - 1}{-1 - 2}\right)(x - 2)$

$y - 1 = \frac{-2}{3}(x - 2)$

$2x + 3y - 7 = 0$

b. $y - 0 = \left(\frac{2 - 0}{4 - 3}\right)(x - 3)$

$y = 2(x - 3)$

$y = 2x - 6$

$2x - y - 6 = 0$

32. Express the equation in slope intercept form:

$$by = -ax - c$$
$$y = \frac{-a}{b}x - \frac{c}{b}.$$

The slope is $\frac{-a}{b}$.

34. Let m_1 be the slope of the line that is the graph of the given equation and m_2 be the slope of the perpendicular. Then $m_1m_2 = -1$. From Problem 32, $m_1 = \frac{-a}{b}$. Hence,

$$\frac{-a}{b}m_2 = -1$$
$$m_2 = \frac{b}{a}.$$

EXERCISE 7.6

2. First graph the equality $y = x$ with a broken line since $y > x$ is false for the components of every point on the line.

   Since (0,0) is a point on the graph of $y = x$, choose another such as (5,3) and substitute the component 5 for x and the component 3 for y in the inequality $y > x$. Since $3 > 5$ is false, the point associated with (5,3) is *not* in the solution set; the points in the other half-plane are the graphs of the ordered pairs in the solution set. Cross hatch or shade this half-plane.

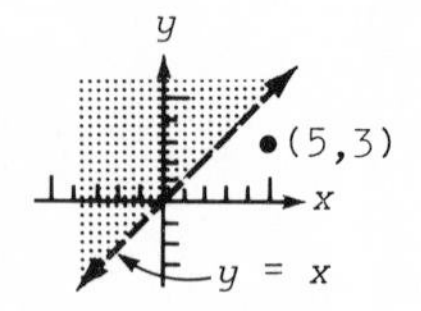

In Problems 4-12, follow the example above and the example on page 238 in the text, remembering to use a broken line when the $<$ or $>$ symbol is used in the given inequality and a solid line when the $\leqslant$ or $\geqslant$ symbol is used.

4.

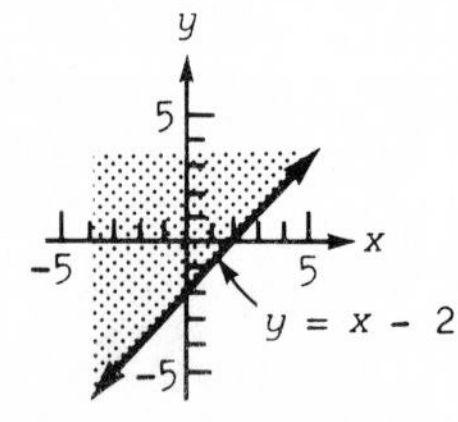

6.

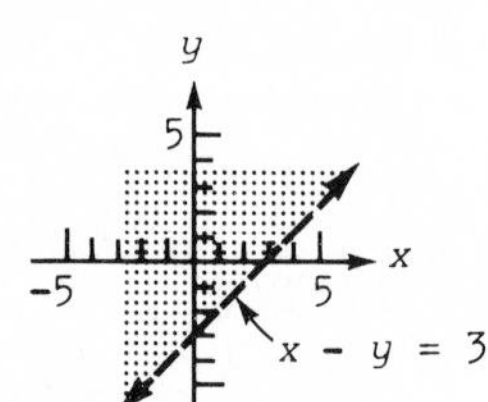

8.

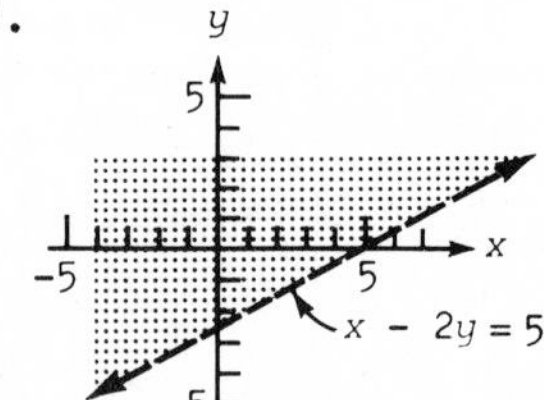

10.

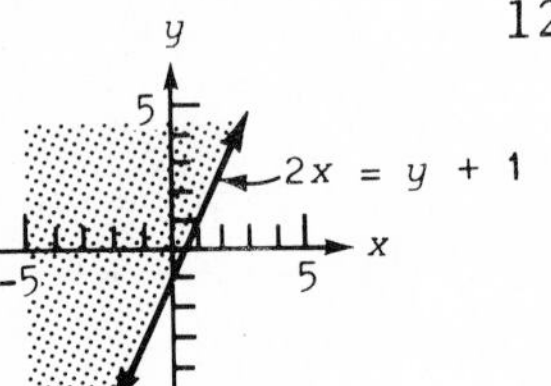

12.

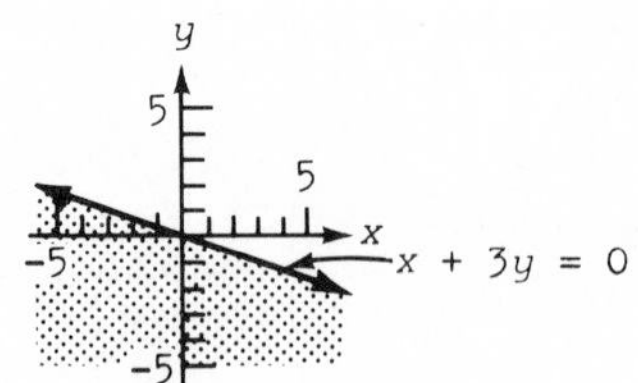

14.

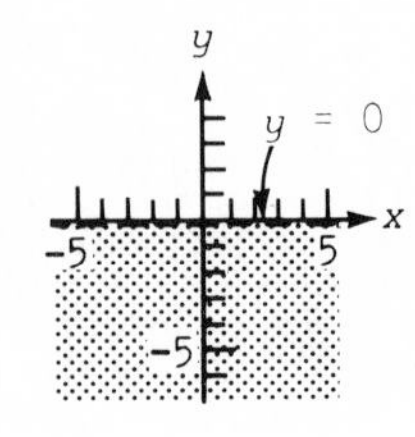

16.

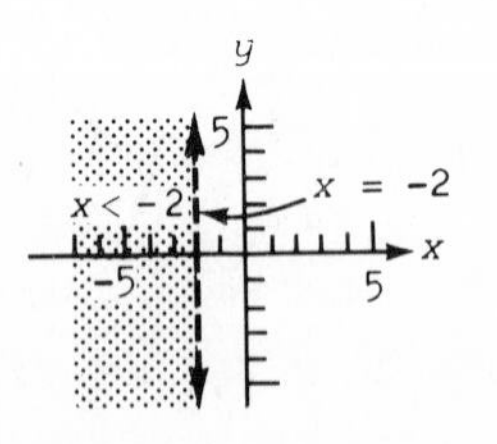

18. $0 \leqslant y \leqslant 1$ means that y is between 0 and 1 including 0 and 1.

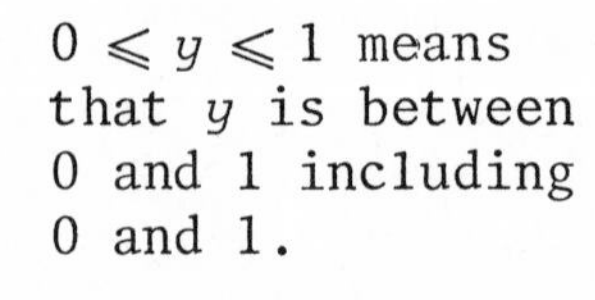

20. Shade the region where x is negative or zero and y is positive or zero.

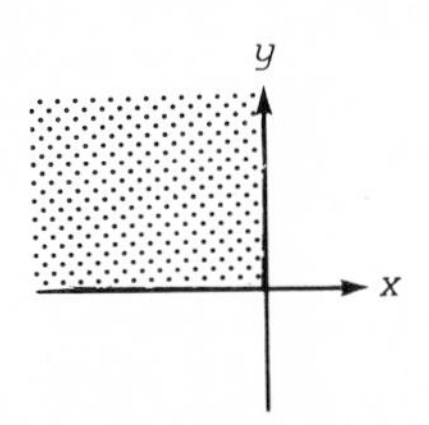

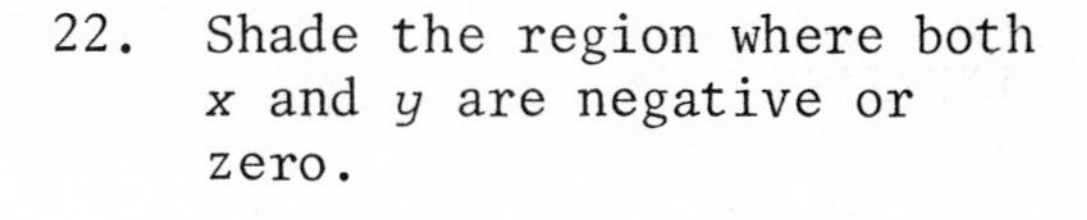

22. Shade the region where both x and y are negative or zero.

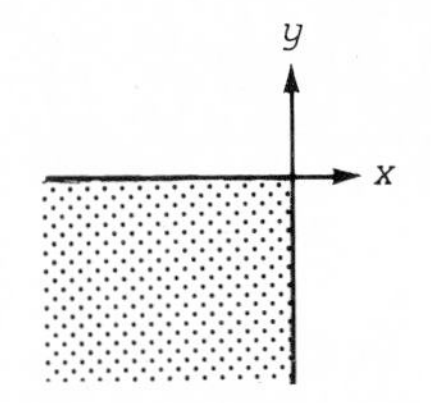

24. Graph $\{(x,y)\,|\,x \leqslant 2\}$. Then graph $\{(x,y)\,|\,y \leqslant 2\}$ with the shading lines running in a direction different from those in the graph of $\{(x,y)\,|\,x \leqslant 2\}$. The required graph will consist of all the points *common* to both shaded regions.

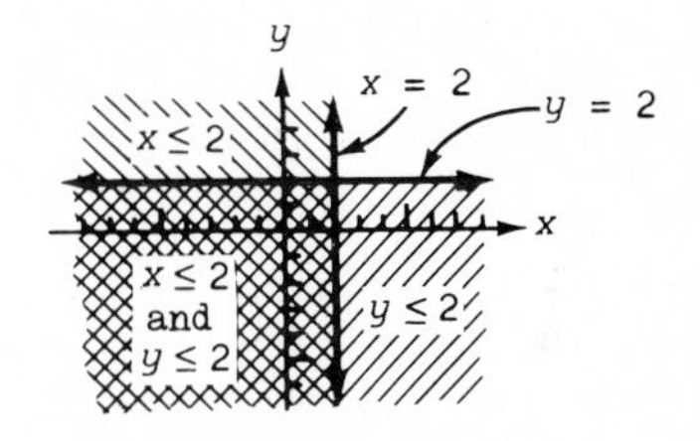

26.

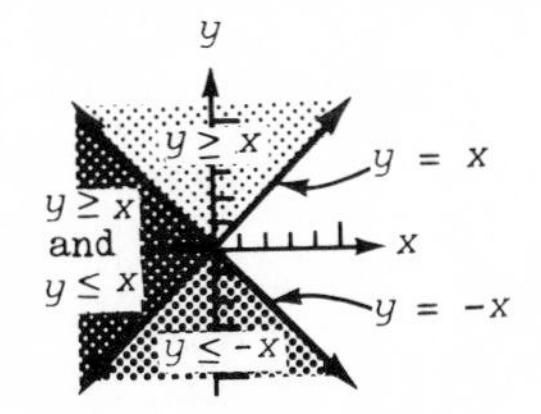

28. $\{(x,y)\,|\,y < |x|\}$ is equivalent to

$$\{(x,y)\,|\,y < x, \quad x \geqslant 0\}$$

or

$$\{(x,y)\,|\,y < -x, \; x < 0\}$$

The required graph is the union of the graphs of each of the above relations.

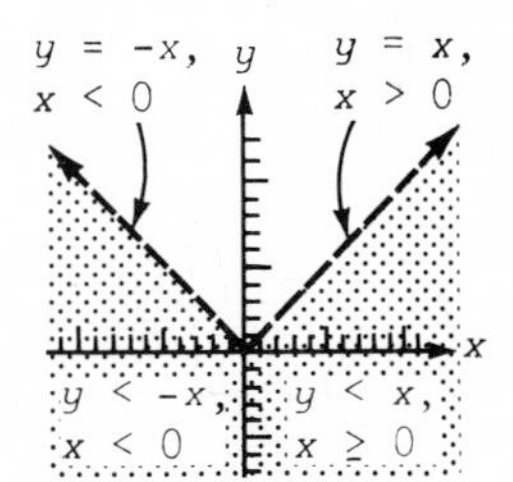

30. $\{(x,y)\,|\,y \geqslant |x| + 5\}$ is equivalent to

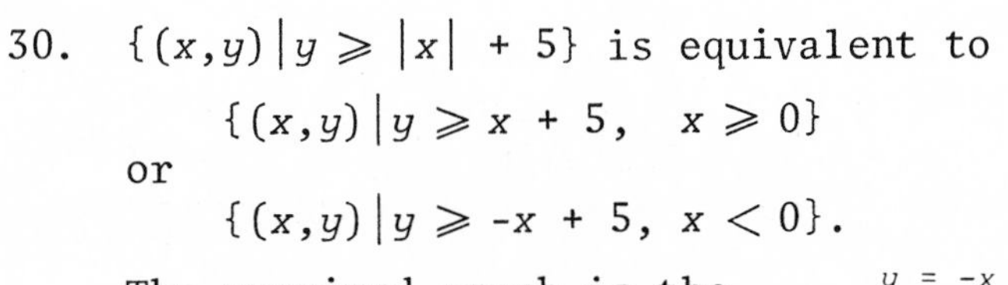

$$\{(x,y)\,|\,y \geqslant x + 5, \quad x \geqslant 0\}$$

or

$$\{(x,y)\,|\,y \geqslant -x + 5, \; x < 0\}.$$

The required graph is the union of graphs of these relations.

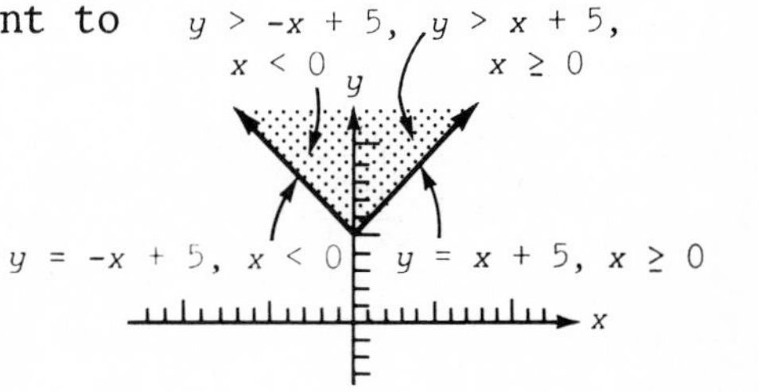

8

FUNCTIONS, RELATIONS, AND THEIR GRAPHS: PART II

EXERCISE 8.1

2. a. Find the ordered pairs corresponding to the integral values of x between -4 and 4. These are (-4,20), (-3,13), (-2,8), (-1,5), (0,4), (1,5), (2,8), (3,13), (4,20).

   b. Graph these ordered pairs and draw a smooth parabola joining them.

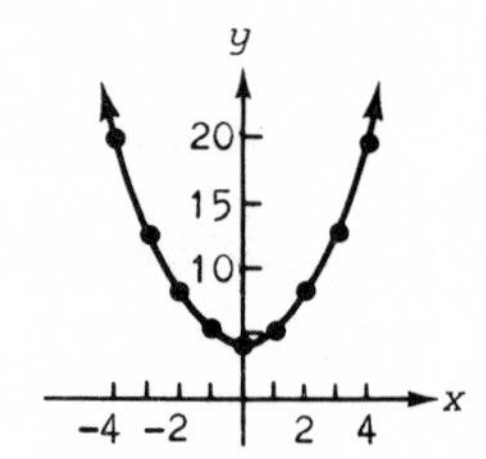

4. a. The solutions are (-4,11), (-3,4), (-2,-1), (-1,-4), (0,-5), (1,-4), (2,-1), (3,4), (4,11).

   b.

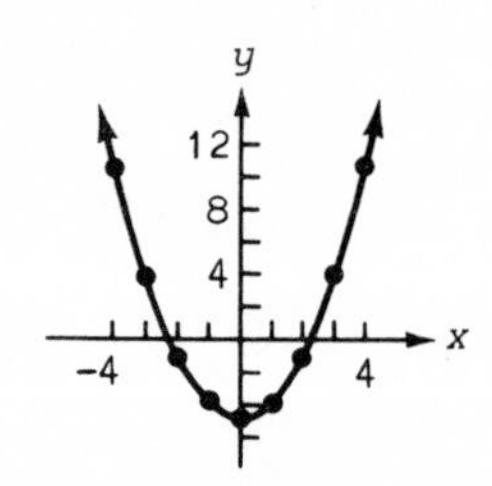

6. a. The solutions are (-4,-11), (-3,-4), (-2,1), (-1,4), (0,5), (1,4), (2,1), (3,-4), (4,-11).

   b.

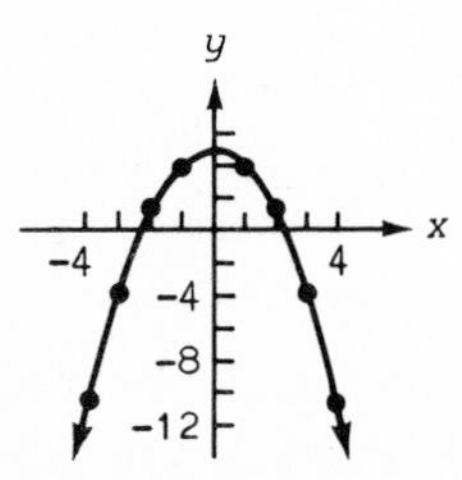

8. a. The solutions are (-4,84), (-3,48), (-2,22), (-1,6) (0,0), (1,4), (2,18), (3,42), (4,76).

b. 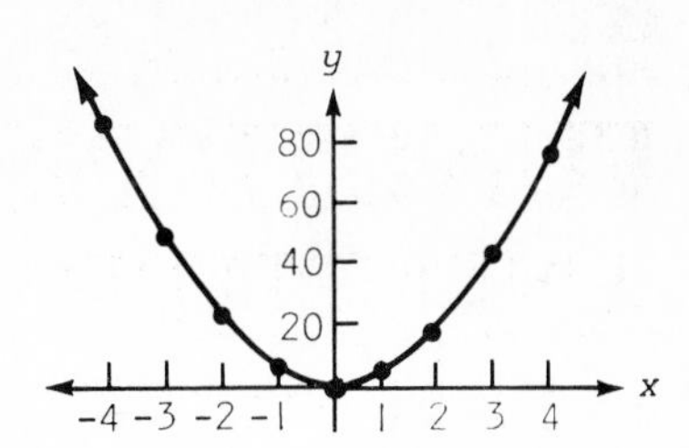

10. a. Write $y = (x - 1)^2$ and then find the values of y. The solutions are (-4,25), (-3,16), (-2,9), (-1,4), (0,1), (1,0), (2,1), (3,4), (4,9).

b. 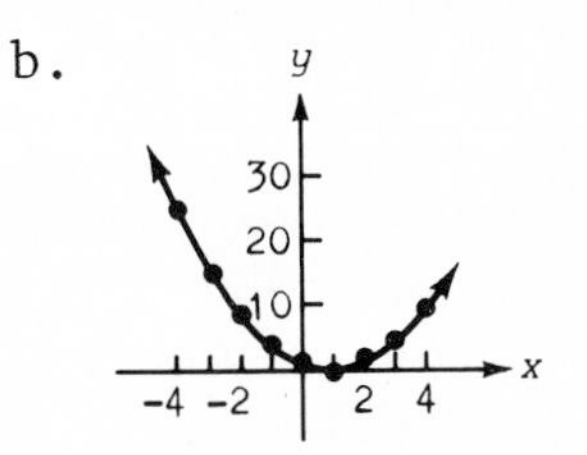

12. a. Write

$$\begin{aligned} f(x) &= -(x^2 - 2x + 1) \\ &= -(x - 1)^2. \end{aligned}$$

Use this last expression to compute the values of y. The solutions are (-4,-25), (-3,-16), (-2,-9), (-1,-4), (0,-1), (1,0), (2,-1), (3,-4), (4,-9).

b. 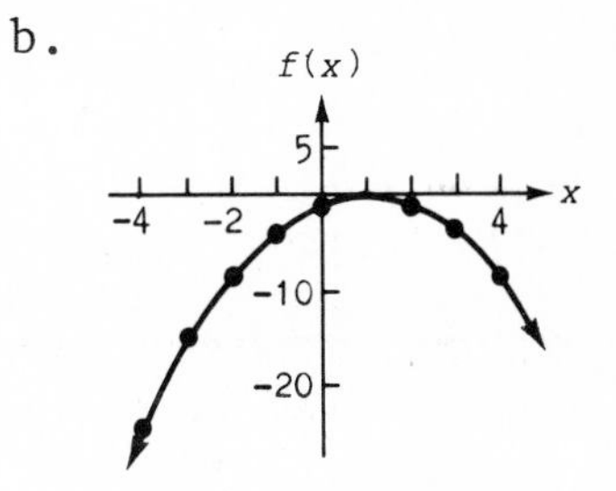

14. a. The solutions are (-4,-54), (-3,-32), (-2,-16), (-1,-6), (0,-2), (1,-4), (2,-12), (3,-26), (4,-46).

b. 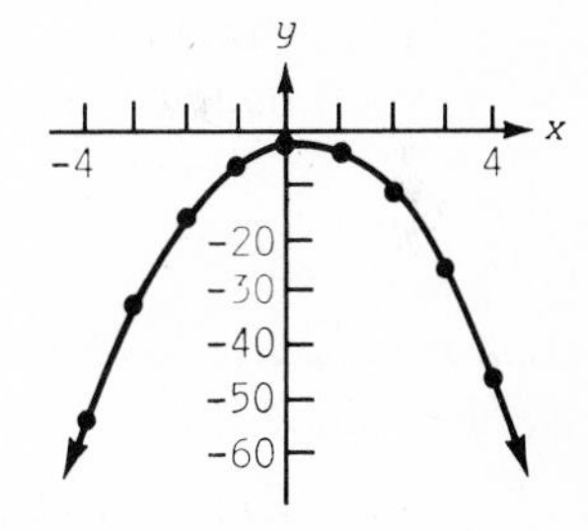

16. a. The solutions are (12,-4), (5,-3), (0,-2), (-3,-1), (-4,0), (-3,1), (0,2), (5,3), (12,4).

b. 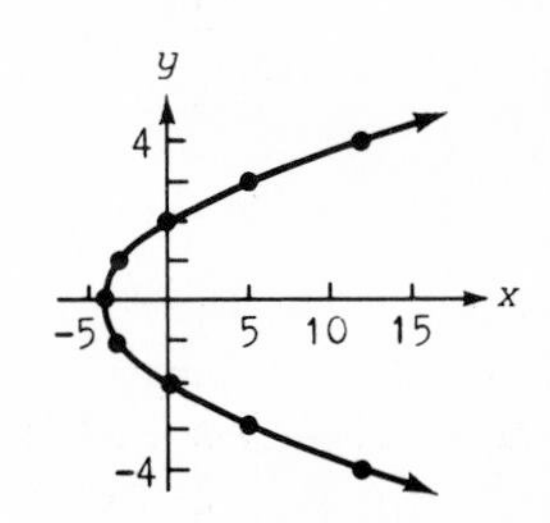

18. a. The solutions are (-38,-4), (-24,-3), (-14,-2), (-8,-1), (-6,0), (-8,1), (-14,2), (-24,3), (-38,4).

b. 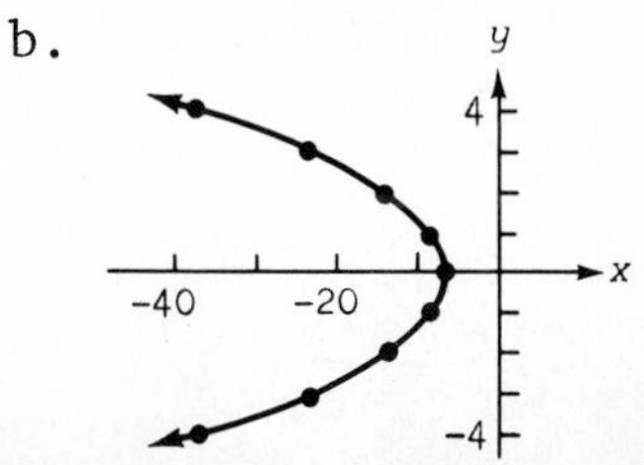

20. a. The solutions are (-18,-4), (-10,-3), (-4,-2), (0,-1), (2,0), (2,1), (0,2), (-4,3), (-10,4).

b.

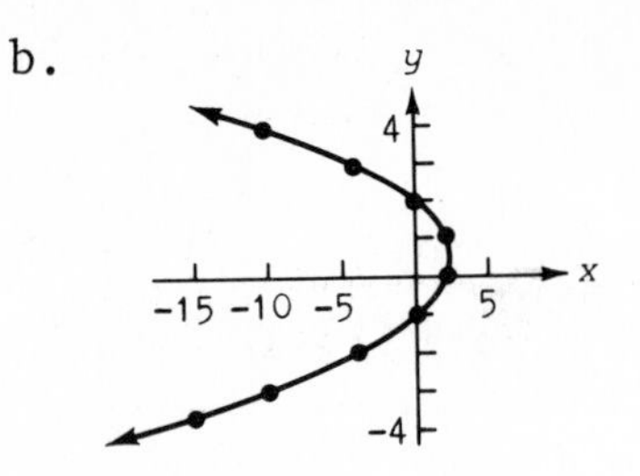

22. Some solutions are (-4,15), (-3,8), (-2,3), (-1,0), (0,-1), (1,0), (2,3), etc. Complete the graph. At $x = -3$ and $x = 2$, vertical segments are drawn from the x-axis terminating at the graph.

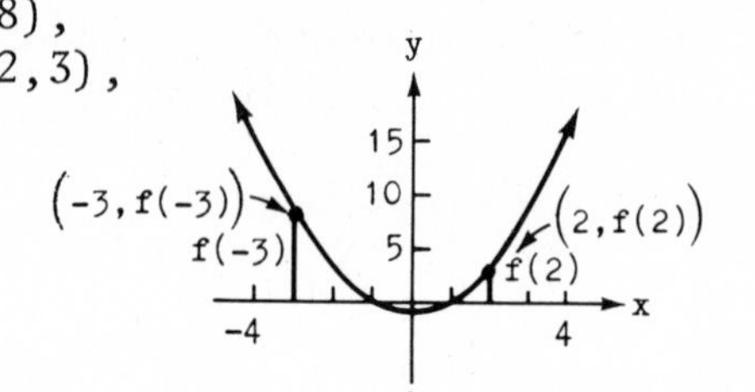

24. a. Replace x by $-x$ and we have $f(-x) = -x$. Since $f(-x) \neq f(x)$, this graph is not symmetric with respect to the y-axis.

b. Here, $f(-x) = (-x)^2 = x^2 = f(x)$. Thus, the graph is symmetric with respect to the y-axis.

c. $f(-x) = \sqrt{(-x)^2 + 1} = \sqrt{x^2 + 1} = f(x)$. Thus, the graph is symmetric with respect to the y-axis.

d. $f(-x) = |-x| = |x| = f(x)$. Thus, the graph is symmetric with respect to the y-axis.

EXERCISE 8.2

2. 1. Since $a = 1 > 0$, the curve opens upward.

2. The x-coordinate of the vertex is

$$-\frac{b}{2a} = -\frac{1}{2(1)} = -\frac{1}{2}.$$

3. Substitute $-\frac{1}{2}$ for x in the equation to obtain the y-coordinate of the vertex:

$$y = \left(-\frac{1}{2}\right)^2 + \left(-\frac{1}{2}\right) - 6 = -\frac{25}{4}.$$

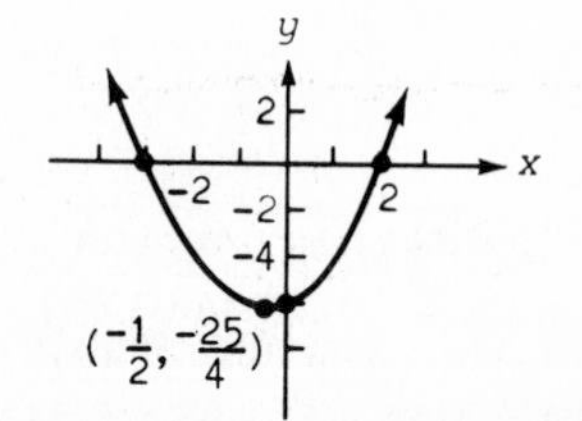

4. If $x = 0$, $y = -6$, so the y-intercept is -6.

If $y = 0$, then

$$x^2 + x - 6 = 0,$$
$$(x + 3)(x - 2) = 0,$$
$$x = -3 \quad \text{or} \quad x = 2,$$

so that the x-intercepts are -3 and 2.

4. Since $a = 1 > 0$, the curve opens upward.

The x-coordinate of the vertex is $-\dfrac{6}{2(1)} = -3.$

The second coordinate of the vertex is
$f(-3) = (-3)^2 + 6(-3) = -9.$
If $x = 0$, $f(x) = 0$. The $f(x)$-intercept is 0.

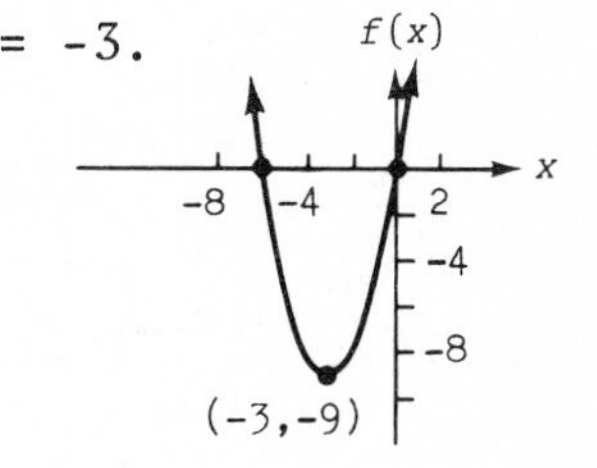

Replace $f(x)$ by 0 and solve the equation.

$$x^2 + 6x = 0$$
$$x(x + 6) = 0$$

The solution set is $\{0,-6\}$; therefore, the x-intercepts are 0 and -6.

6. Since $a = 1 > 0$, the curve opens upward.

The x-coordinate of the vertex is
$-\dfrac{b}{2a} = -\dfrac{6}{2(1)} = -3.$ The second coordinate is

$$g(-3) = (-3)^2 + 6(-3) + 8 = -1.$$

Replace $g(x)$ by 0 and solve the equation.

$$x^2 + 6x + 8 = 0$$
$$(x + 4)(x + 2) = 0$$

The solution set is $\{-4,-2\}$; therefore, the x-intercepts are -4 and -2. Replace x by 0 and obtain $g(x) = 8$.
The $g(x)$-intercept is 8.

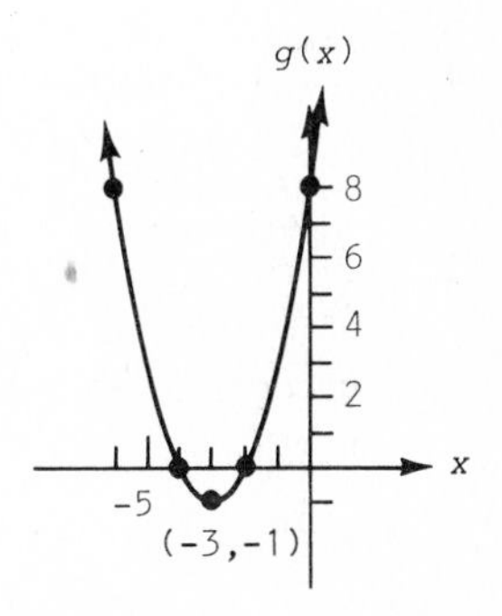

8. Since $a = 3 > 0$, the curve opens upward.

The x-coordinate of the vertex is
$-\dfrac{-5}{2(3)} = \dfrac{5}{6}.$

The y-coordinate of the vertex is

$$y = 3\left(\frac{5}{6}\right)^2 - 5\left(\frac{5}{6}\right) - 2 = \frac{25}{12} - \frac{50}{12} - \frac{24}{12} = -\frac{49}{12}.$$

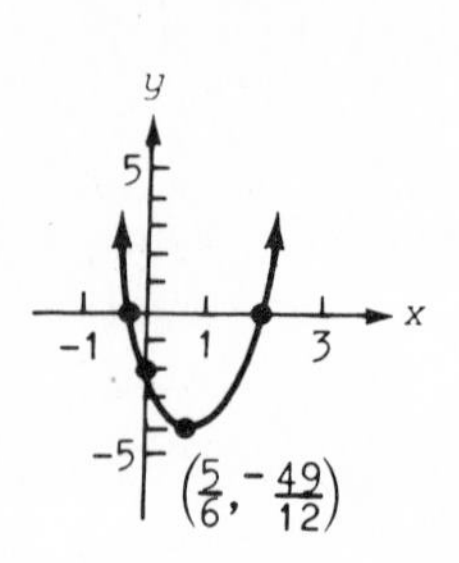

Solving $3x^2 - 5x - 2 = 0$, we have

$$(3x + 1)(x - 2) = 0,$$

$$x = \frac{-1}{3} \quad \text{or} \quad x = 2.$$

The x-intercepts are -1/3 and 2. The y-intercept is -2.

10. Since $a = -1 < 0$, the curve opens downward. The x-coordinate of the vertex is

$$-\frac{-4}{2(-1)} = -2.$$

The second coordinate of the vertex is $f(-2) = -(-2)^2 - 4(-2) + 5 = 9$.

Replace $f(x)$ by 0 and solve the equation.

$$-(x^2 + 4x - 5) = 0$$
$$-(x + 5)(x - 1) = 0$$

The solution set is $\{-5,1\}$ and the x-intercepts are -5 and 1. Replace x by 0 and obtain $f(x) = 5$. The $f(x)$-intercept is 5.

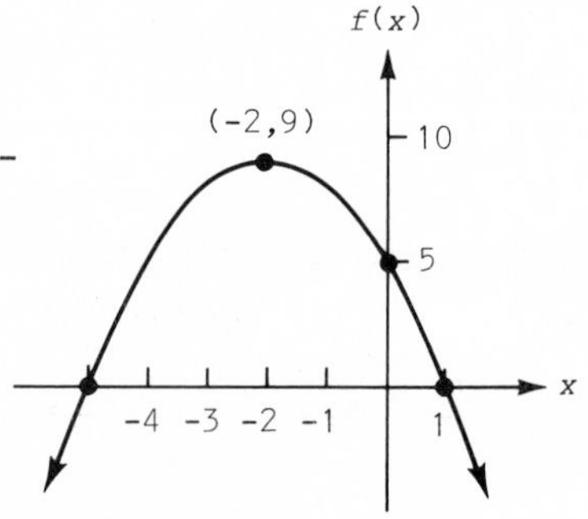

12. Since $a = -3 < 0$, the curve opens downward. The x-coordinate of the vertex is

$$-\frac{7}{2(-3)} = \frac{7}{6}.$$

The y-coordinate of the vertex is

$$y = -3\left(\frac{7}{6}\right)^2 + 7\left(\frac{7}{6}\right) - 2 = -\frac{49}{12} + \frac{98}{12} - \frac{24}{12}$$
$$= \frac{25}{12}.$$

Replace y by 0 and solve the equation.

$$-(3x^2 - 7x + 2) = 0$$
$$-(3x - 1)(x - 2) = 0$$

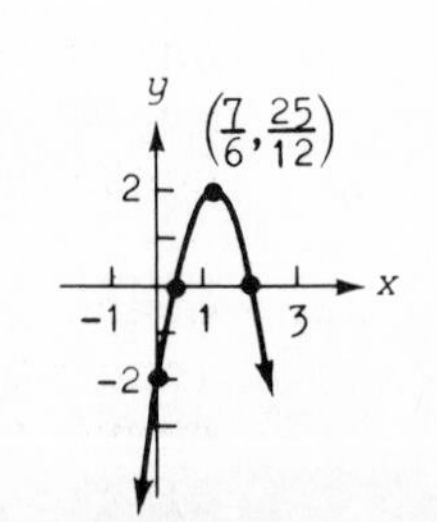

The solution set is $\left\{\frac{1}{3}, 2\right\}$; therefore, the x-intercepts are $\frac{1}{3}$ and 2. Replace x by 0 and obtain $y = -2$. The y-intercept is -2.

14. $A = x(50 - x)$; $A = -x^2 + 50x$. The x-coordinate of the vertex is

$$-\frac{50}{2(-1)} = 25.$$

Hence, $x = 25$ yields the maximum value for A which is

$$25(50 - 25) = 25(25) = 625.$$

The maximum area is 625 square inches.

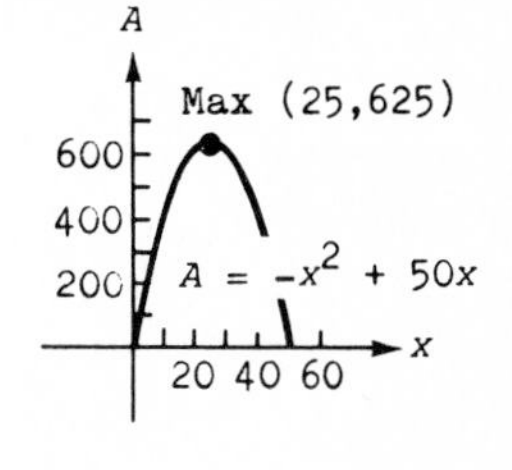

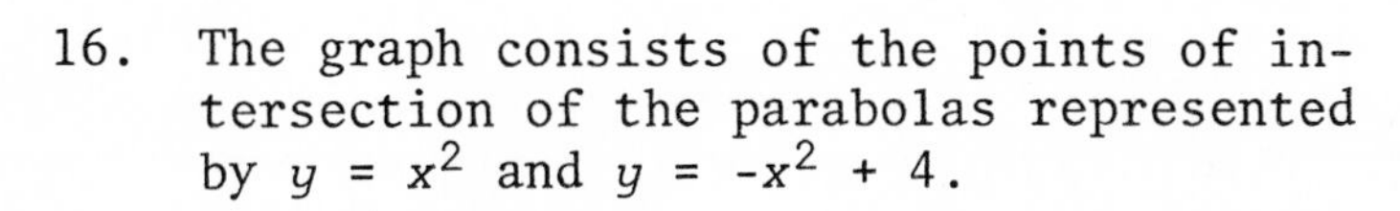
16. The graph consists of the points of intersection of the parabolas represented by $y = x^2$ and $y = -x^2 + 4$.

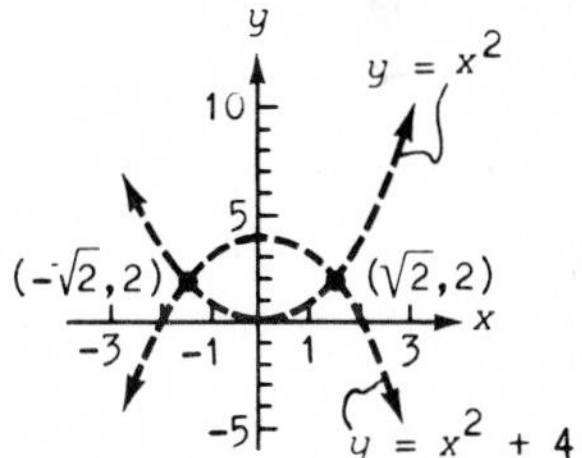

18. The graph consists of all points on each of the parabolas represented by $y = x^2 + 1$ and $y = -x^2 - 1$.

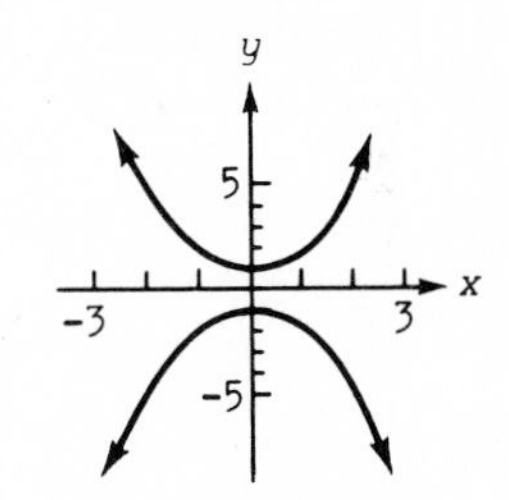

20. Graph $y = x^2 - 6x + 8$ using a solid line to indicate that points on the curve are included in the graph of the given inequality. Arbitrarily selecting (0,0) and substituting into $y \leqslant x^2 - 6x + 8$, we obtain

$$0 \leqslant 0^2 - 6(0) + 8,$$

which is true. Hence, we shade the region below the parabola because it contains (0,0)

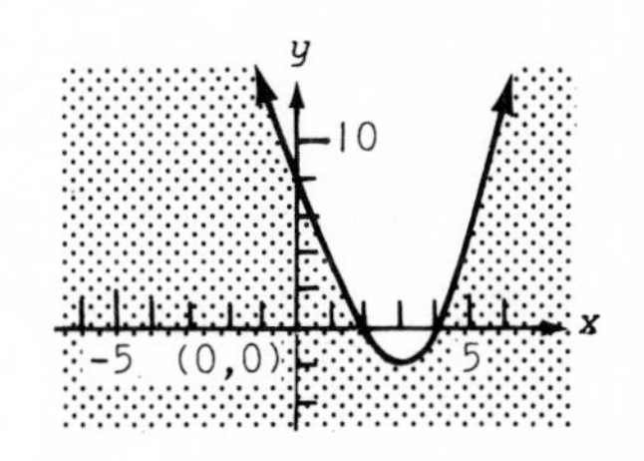

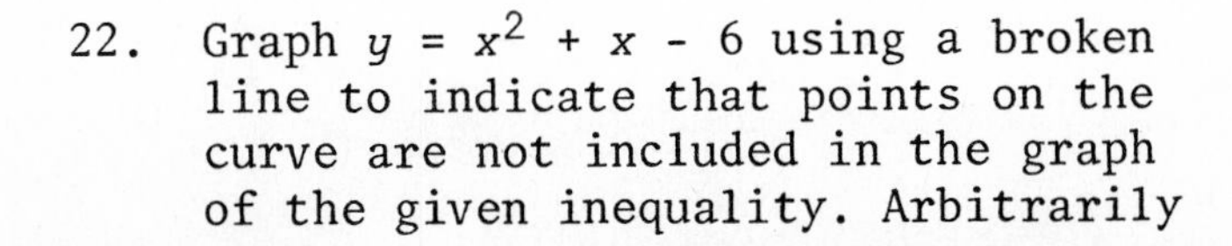
22. Graph $y = x^2 + x - 6$ using a broken line to indicate that points on the curve are not included in the graph of the given inequality. Arbitrarily

selecting (0,0) and substituting into $y > x^2 + x - 6$, we obtain

$$0 > 0^2 + 0 - 6,$$

which is true. Hence, we shade the region above the parabola because it contains (0,0).

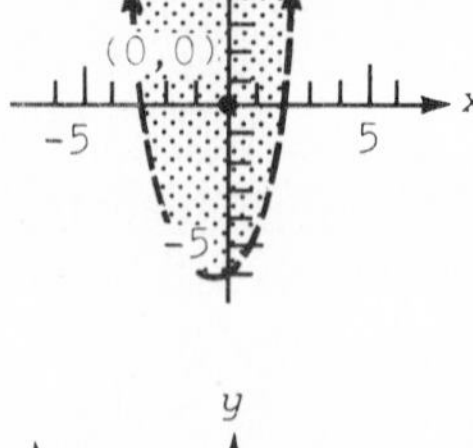

24. The graph consists of all points on or above the parabola represented by $y = x^2 + 3x + 2$.

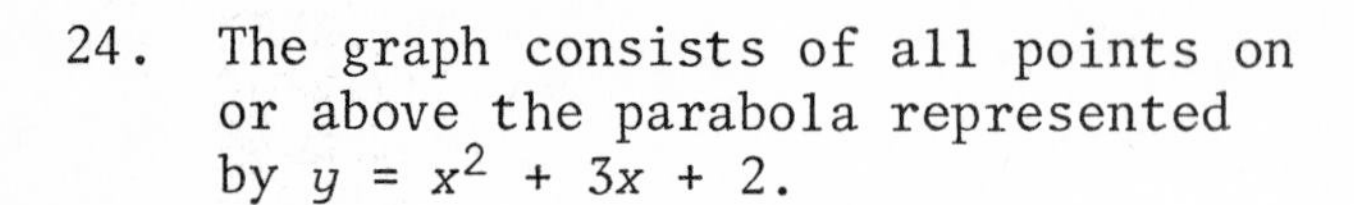

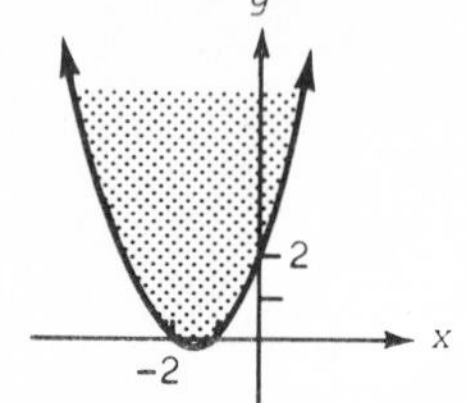

26. The graph consists of all points that are on or below the graph of $y = 1 - x^2$, or that are on or above the graph of $y = -1$.

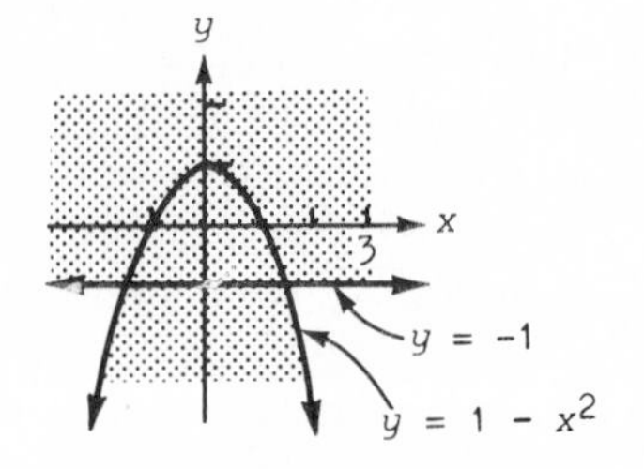

28. The graph consists of all points that are on or below the graph of $y = 4 - x^2$, and, at the same time, on or above the graph of $y = x^2 - 4$.

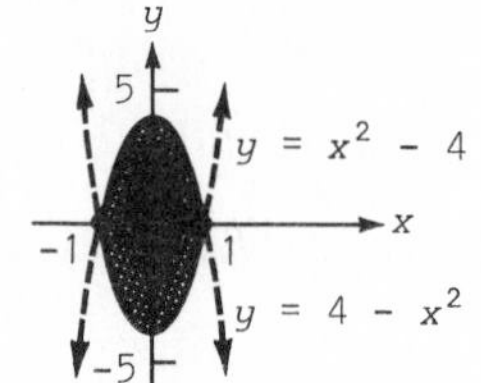

30. For $k = -1$, $y = -x^2$;
$k = -2$, $y = -2x^2$;
$k = -3$, $y = -3x^2$;
$k = -4$, $y = -4x^2$.

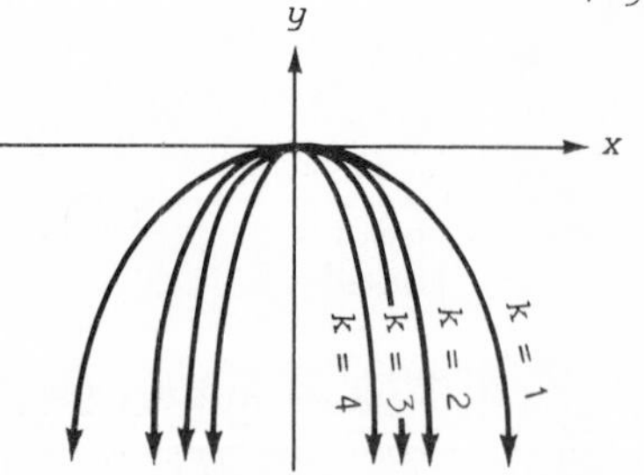

32. For $k = -2$, $y = x^2 - 2x$;
$k = 0$, $y = x^2$;
$k = 2$, $y = x^2 + 2x$;
$k = 4$, $y = x^2 + 4x$;

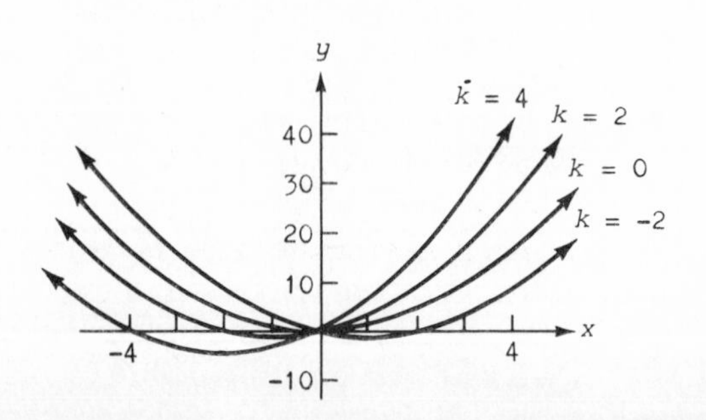

34. The object will strike the ground when $d = 0$. Hence, substituting 0 for d,

$$32t - 8t^2 = 0$$
$$8t(4 - t) = 0.$$

Therefore, $t = 0$ or $t = 4$. This means that the object was at ground level ($d = 0$) at 0 seconds and again at 4 seconds. Hence, the object was in the air for 4 seconds.

36. The graph must contain the points whose coordinates are

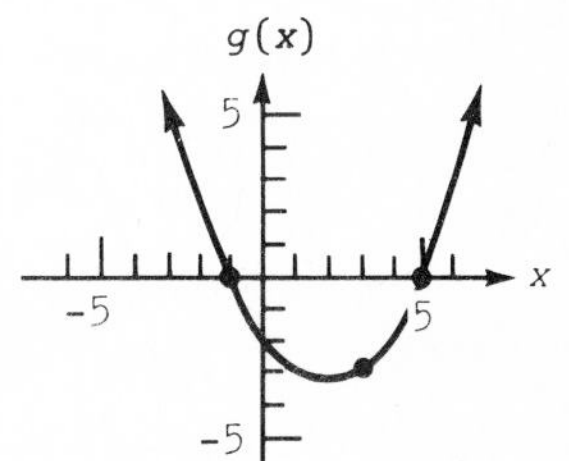

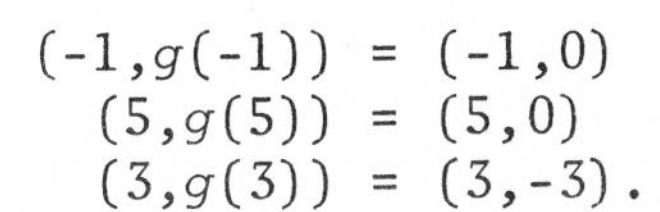

$$(-1,g(-1)) = (-1,0)$$
$$(5,g(5)) = (5,0)$$
$$(3,g(3)) = (3,-3).$$

We must sketch a parabola through these points, since every quadratic function has a parabola for its graph. Furthermore, the graph must be below the x-axis for $-1 < x < 5$ because $g(x) < 0$ for these values.

EXERCISE 8.3

2. a. $y^2 = 9 - x^2$; $y = \pm\sqrt{9 - x^2}$

   b. y will be a real number if $9 - x^2 \geqslant 0$. Solving this inequality, we find that the domain is $\{x | -3 \leqslant x \leqslant 3\}$.

   c. Some solutions are $(-3,0)$, $(-2,\pm\sqrt{5})$, $(-1,\pm 2\sqrt{2})$, $(0,\pm 3)$, $(1,\pm 2\sqrt{2})$, $(2,\pm\sqrt{5})$, $(3,0)$.

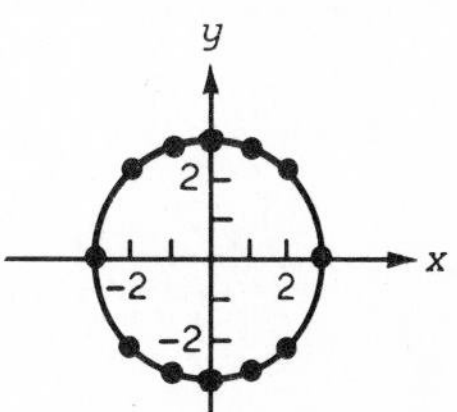

4. a. $y^2 = 4 - 4x^2$; $y = \pm\sqrt{4 - 4x^2} = \pm 2\sqrt{1 - x^2}$

   b. y will be a real number if $1 - x^2 \geqslant 0$. Solving this inequality, we find that the domain is $\{x | -1 \leqslant x \leqslant 1\}$.

   c. Some solutions are $(-1,0)$, $\left(-\frac{1}{2},\pm\sqrt{3}\right)$, $(0,\pm 2)$, $\left(\frac{1}{2},\pm\sqrt{3}\right)$, $(1,0)$.

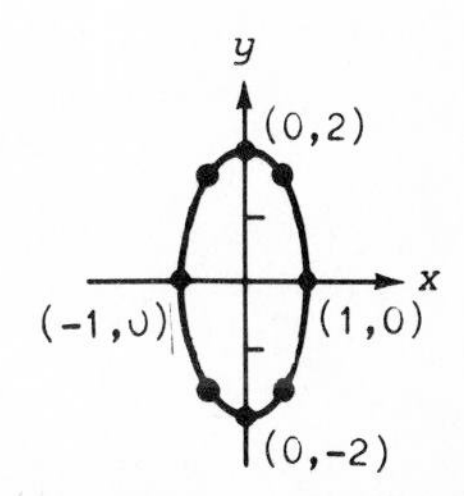

6. a. $9y^2 = 4 - x^2$; $y = \frac{\pm\sqrt{4 - x^2}}{3}$

b. Solve $4 - x^2 \geqslant 0$. The domain is $\{x | -2 \leqslant x \leqslant 2\}$.

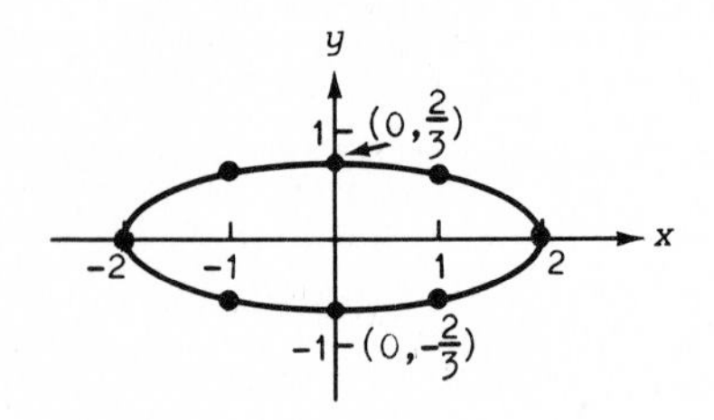

c. Some solutions are $(-2,0)$ $\left(-1,\pm\frac{\sqrt{3}}{3}\right)$, $\left(0,\pm\frac{2}{3}\right)$, $\left(1,\pm\frac{\sqrt{3}}{3}\right)$, $(2,0)$.

8. a. $-y^2 = 1 - 4x^2$; $y^2 = 4x^2 - 1$;
$y = \pm\sqrt{4x^2 - 1}$

b. Solving $4x^2 - 1 \geqslant 0$, we have $x \geqslant \frac{1}{2}$ or $x \leqslant \frac{-1}{2}$. Therefore, the domain is $\left\{x | x \leqslant \frac{-1}{2} \text{ or } x \geqslant \frac{1}{2}\right\}$.

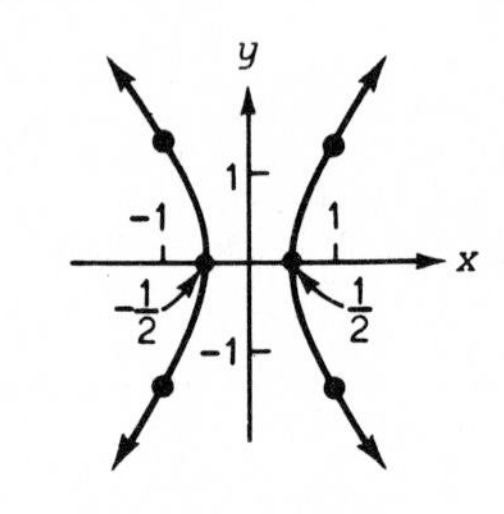

c. Some solutions are $(-1,\pm\sqrt{3})$, $\left(-\frac{1}{2},0\right)$, $\left(\frac{1}{2},0\right)$, $(1,\pm\sqrt{3})$.

10. a. $y^2 = \frac{36 + 9x^2}{4}$; $y = \frac{\pm\sqrt{9(4 + x^2)}}{2}$;
$y = \frac{\pm 3\sqrt{4 + x^2}}{2}$

b. Since $4 + x^2 \geqslant 0$ for all real numbers, the domain is the set of all real numbers

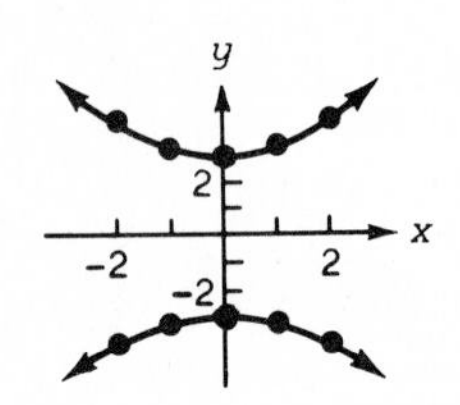

c. Some solutions are $(-2,\pm 3\sqrt{2})$, $\left(-1,\pm\frac{3\sqrt{5}}{2}\right)$, $(0,\pm 3)$, $\left(1,\pm\frac{3\sqrt{5}}{2}\right)$, $(2,\pm 3\sqrt{2})$.

12. a. $3y^2 = 12 - 4x^2$; $y^2 = \frac{4(3 - x^2)}{3}$;
$y = \pm 2\sqrt{\frac{3 - x^2}{3}} = \pm\frac{2}{3}\sqrt{9 - 3x^2}$

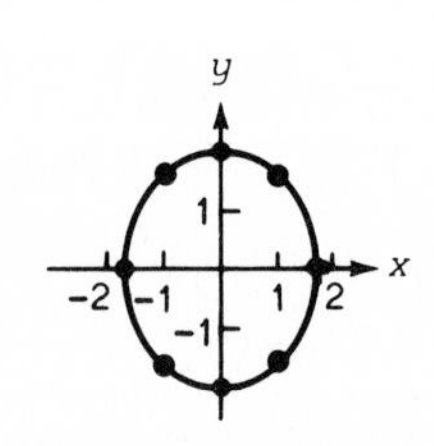

b. Solve $3 - x^2 \geqslant 0$. The domain is $\{x | -\sqrt{3} \leqslant x \leqslant \sqrt{3}\}$.

c. Some solutions are $(-\sqrt{3},0)$, $\left(-1,\pm\frac{2}{3}\sqrt{6}\right)$, $(0,\pm 2)$, $\left(1,\pm\frac{2}{3}\sqrt{6}\right)$, $(\sqrt{3},0)$.

14. $y = \dfrac{-4}{x}$; some solutions are

$(-4,1)$, $\left(-3,\frac{4}{3}\right)$, $(-2,2)$, $(-1,4)$,

$(1,-4)$, $(2,-2)$, $\left(3,\frac{-4}{3}\right)$, $(4,-1)$.

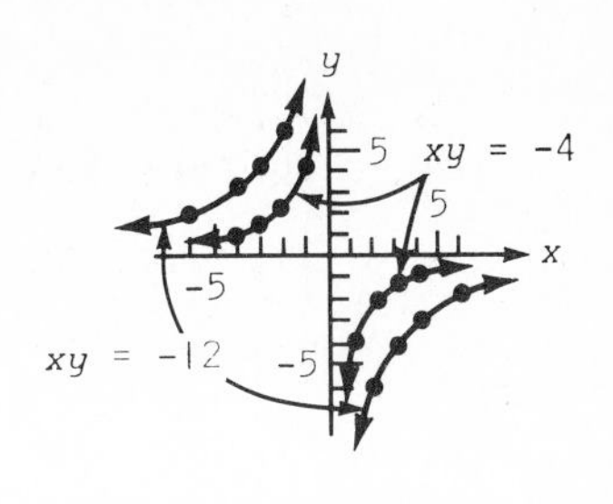

$y = \dfrac{-12}{x}$; some solutions are

$(-6,2)$, $(-4,3)$, $(-3,4)$, $(-2,6)$.

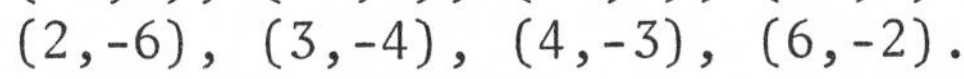
$(2,-6)$, $(3,-4)$, $(4,-3)$, $(6,-2)$.

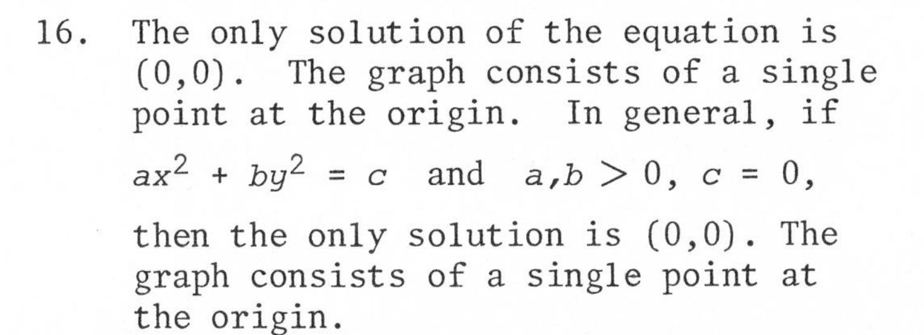
16. The only solution of the equation is $(0,0)$. The graph consists of a single point at the origin. In general, if

$ax^2 + by^2 = c$ and $a,b > 0$, $c = 0$,

then the only solution is $(0,0)$. The graph consists of a single point at the origin.

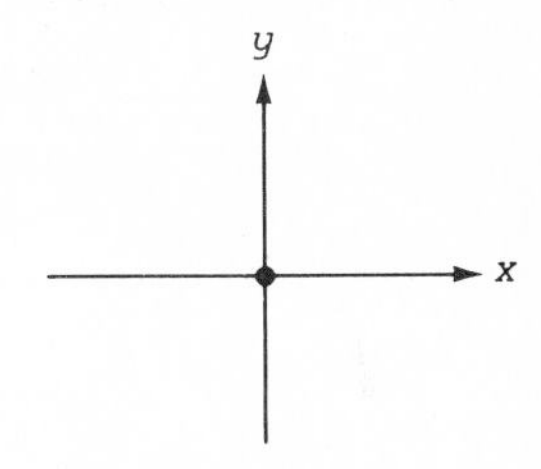

18. Let (x,y) be the coordinates of all the points whose distances from the graph of (h,k) equal r. Then

$$\sqrt{(x - h)^2 + (y - k)^2} = r.$$

Squaring both members yields

$$(x - h)^2 + (y - k)^2 = r^2.$$

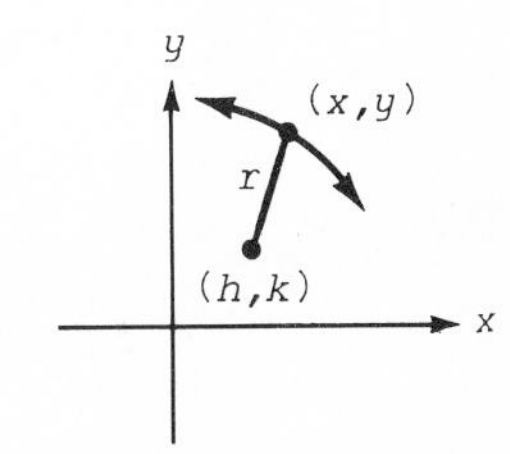

EXERCISE 8.4

2. a. $x^2 + y^2 = 8^2$; a, b, and c have like signs and $a = b$. The graph is a circle.

   b. The x- and y-intercepts are ± 8.

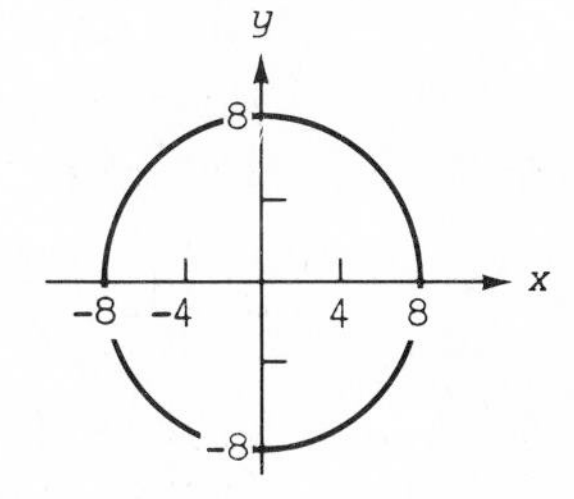

4. a. a, b, $c > 0$, $a \neq b$; the graph is an ellipse.

   b. The x-intercepts are $\pm 2\sqrt{2}$. The y-intercepts are ± 2.

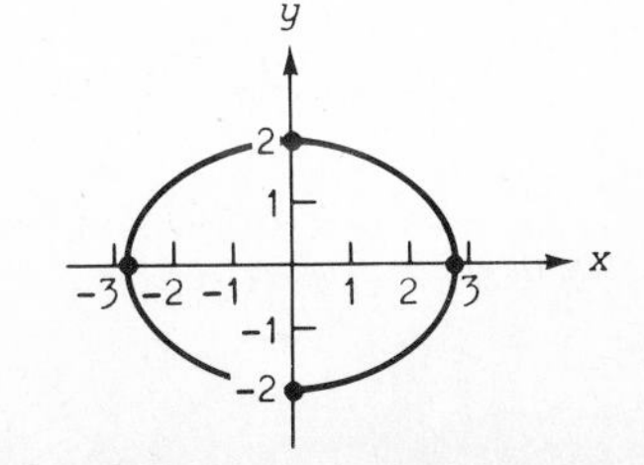

6. a. a and b have opposite signs and $c \neq 0$. The graph is a hyperbola.

   b. There are x-intercepts at $\pm 2\sqrt{2}$. A few other solutions are $(-4,\pm 2)$, $(4,\pm 2)$.

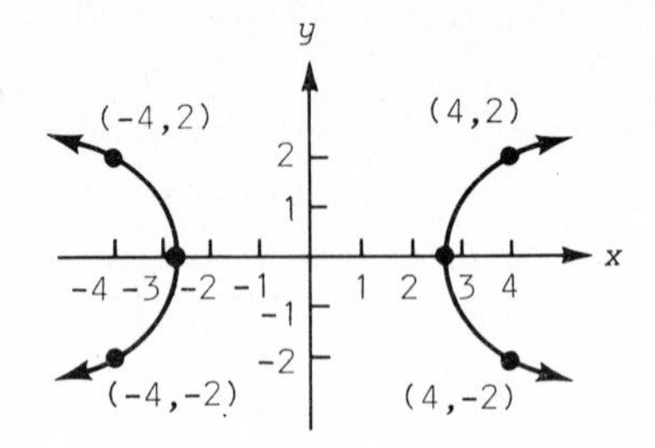

8. a. a and b have opposite signs and $c = 0$. The graph is two distinct lines through the origin.

   b. The equation can be written as

$$(x + 3y)(x - 3y) = 0.$$

   Graph the equations $x + 3y = 0$ and $x - 3y = 0$.

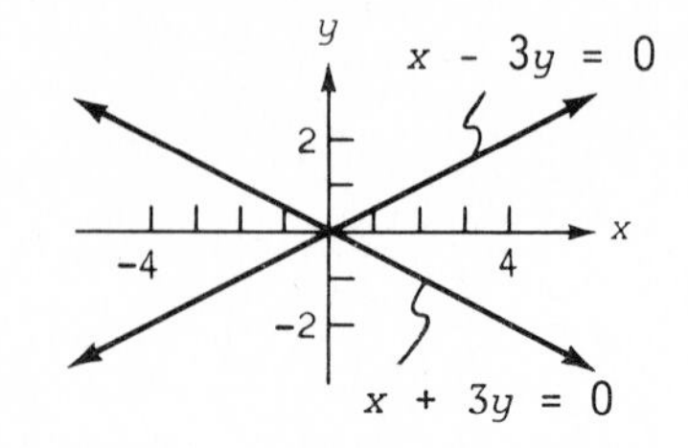

10. a. a and b have the same sign and $c = 0$. The graph is a point.

   b. The only solution is $(0,0)$. Hence, the origin is the graph.

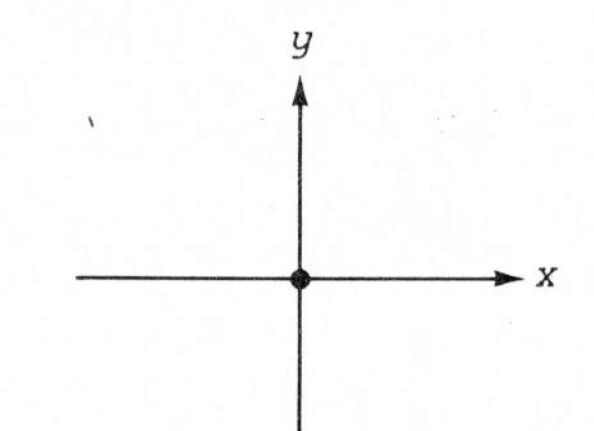

12. a. In $ax^2 + by^2 = c$ form, the equation is $9x^2 + 9y^2 = 2$. a, b, and c have the same sign and $a = b$. The graph is a cirle.

   b. The x- and y-intercepts are $\pm\frac{\sqrt{2}}{3} \approx \pm 0.5$.

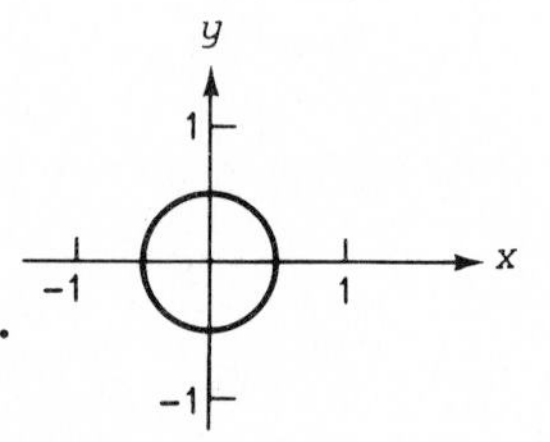

14. a. In $ax^2 + by^2 = c$ form, the equation is $4x^2 + 3y^2 = 12$. a, b, and c have the same sign and $a \neq b$. The graph is an ellipse.

   b. The x-intercepts are $\pm\sqrt{3}$. The y-intercepts are ± 2.

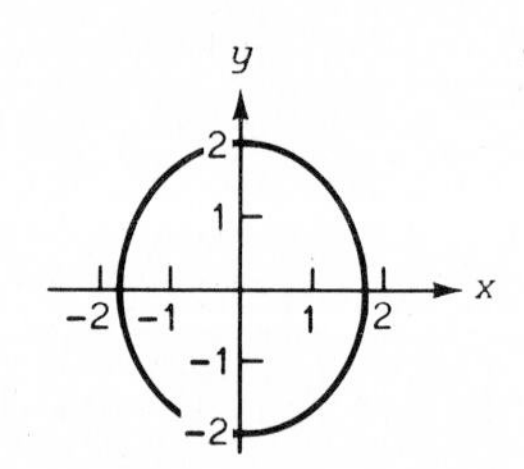

16. a. In $ax^2 + by^2 = c$ form, the equation is $x^2 - 5y^2 = 25$. a and b have opposite signs and $c \neq 0$. The graph is a hyperbola.

   b. There are x-intercepts at ± 5. A few other solutions are

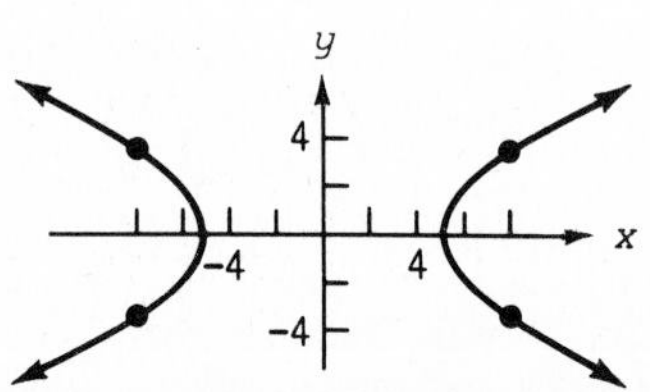

$\left(-8, \pm\sqrt{\frac{39}{5}}\right)$, and $\left(8, \pm\sqrt{\frac{39}{5}}\right)$,
or approximately (-8,±2.8) and (8,±2.8).
Asymptotes are not convenient to use.

18. a. In $ax^2 + by^2 = c$ form, the equation is $4x^2 + y^2 = -6$. $a, b > 0$ and $c < 0$. There is no graph.

20. a. $\frac{9x^2}{4} + \frac{y^2}{4} = 1$; then divide the numerator and denominator of $\frac{9x^2}{4}$ by 9 and obtain $\frac{x^2}{4/9} + \frac{y^2}{4} = 1$.

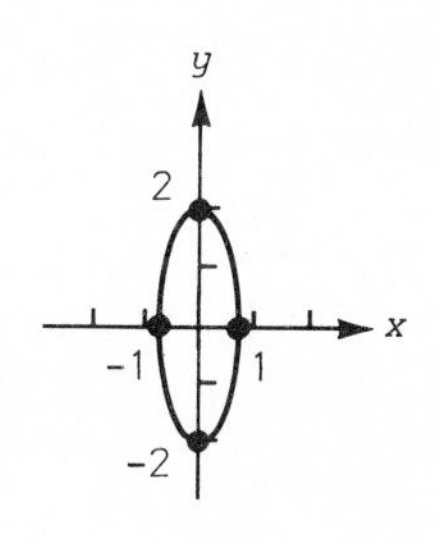

b. The given equation is in the form $ax^2 + by^2 = c$ with a, b, and c having the same sign and $a \neq b$. Therefore, the graph is an ellipse. From standard form, the x-intercepts are $\pm\frac{2}{3}$ and y-intercepts are ±2.

22. a. $\frac{x^2}{9} - \frac{4y^2}{9} = 1$; $\frac{x^2}{9} - \frac{y^2}{9/4} = 1$

b. From the given equation, a and b have opposite signs and $c \neq 0$. Therefore, the graph is a hyperbola. From standard form, the x-intercepts are ±3 and there are no y-intercepts. An equation of the asymptotes is

$$x^2 - 4y^2 = 0; \quad y = \pm\frac{1}{2}x.$$

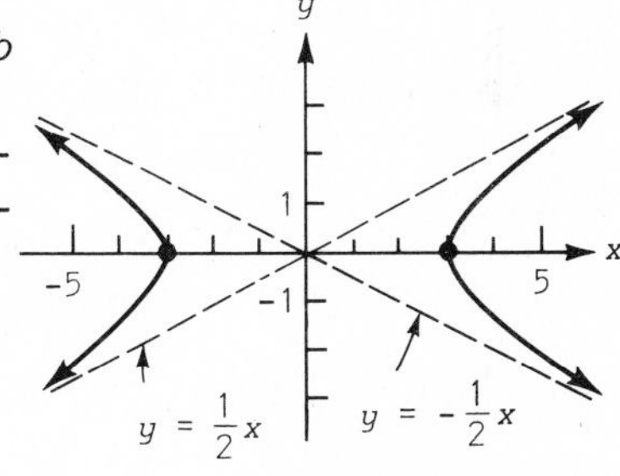

24. a. $\frac{16x^2}{100} + \frac{25y^2}{100} = 1$; $\frac{4x^2}{25} + \frac{y^2}{4} = 1$;
$\frac{x^2}{25/4} + \frac{y^2}{4} = 1$

b. From the given equation, a, b, and c have the same sign and $a \neq b$. Therefore, the graph is an ellipse. From standard form, the x-intercepts are $\pm\frac{5}{2}$ and the y-intercepts are ±2.

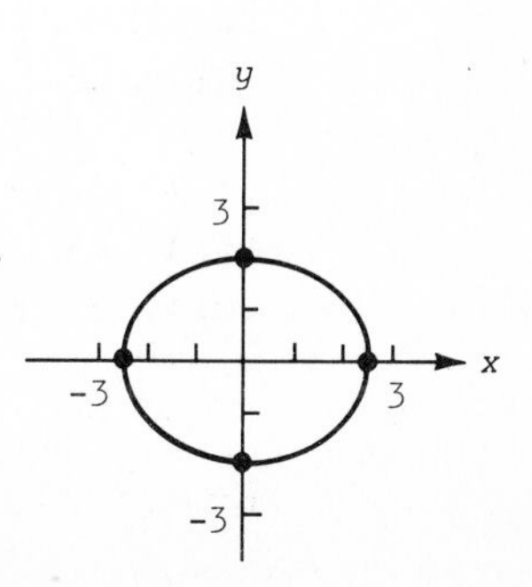

26. a. $\frac{y^2}{4} - \frac{x^2}{10} = 1$

b. From the given equation, a and b have opposite signs and $c \neq 0$. Therefore, the graph is a hyperbola. From standard form, there are y-intercepts at ± 2 and no x-intercepts. An equation of the asymptotes is

$25y^2 - 10x^2 = 0$; $y = \pm\sqrt{\frac{2}{5}}x$, which is not convenient to use. A few arbitrary solutions are obtained by choosing $y = \pm 4$ from which $x = \pm\sqrt{30} \approx \pm 5.5$. Hence, $(\pm 5.5, 4)$ and $(\pm 5.5, -4)$ are points on the hyperbola.

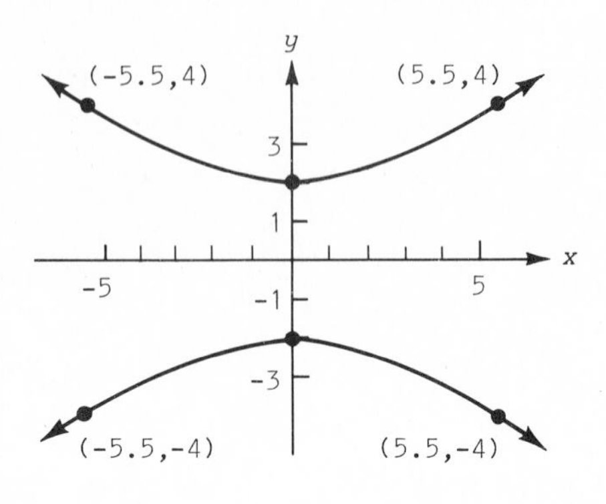

28. $x^2 + y^2 = 16$; a circle of radius 4.

$y = x^2 - 4$; a parabola with x-intercepts ± 2, y-intercept -4, and minimum point $(0,-4)$.

The graph is the union of the two sets of points.

30. $y = x + 2$; a straight line with x-intercept at -2 and y-intercept at 2.

$4x^2 + y^2 = 36$; an ellipse with x-intercepts at ± 3 and y-intercepts at ± 6.

The graph is the union of the two sets of points.

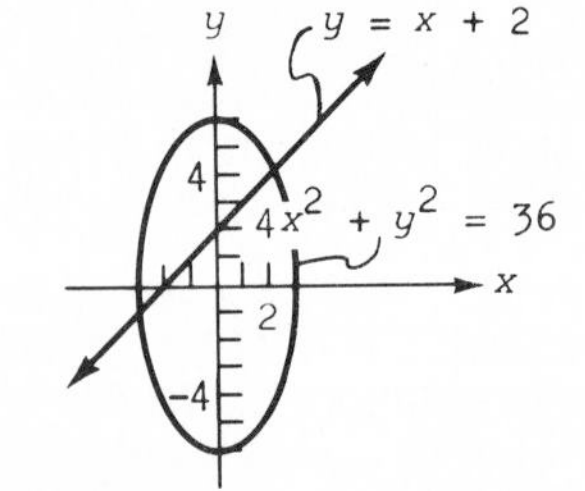

32. Graph $x^2 - y^2 = 1$ with a solid line. Arbitrarily select $(0,0)$ and substitute into $x^2 - y^2 > 1$ and obtain

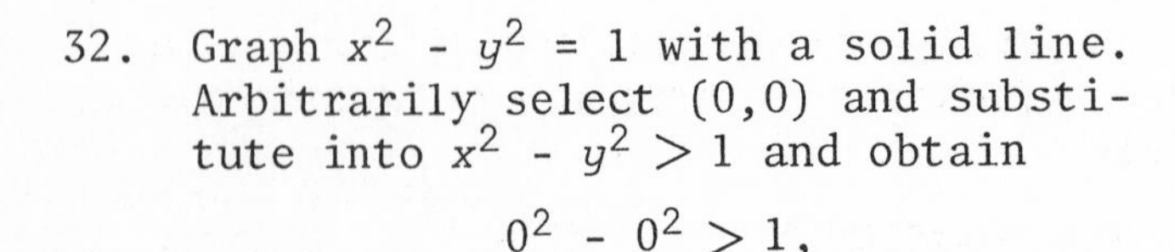

$$0^2 - 0^2 > 1,$$

which is false. Hence, shade the regions not containing $(0,0)$.

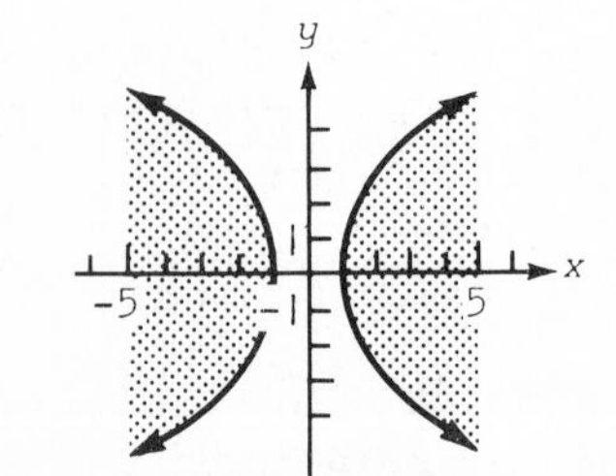

34. Graph $9x^2 - 4y^2 = 36$ with a solid line. Select (0,0) and substitute into $9x^2 - 4y^2 < 36$ and obtain

$$9(0)^2 - 4(0)^2 < 36,$$

which is true. Hence, shade the region containing (0,0).

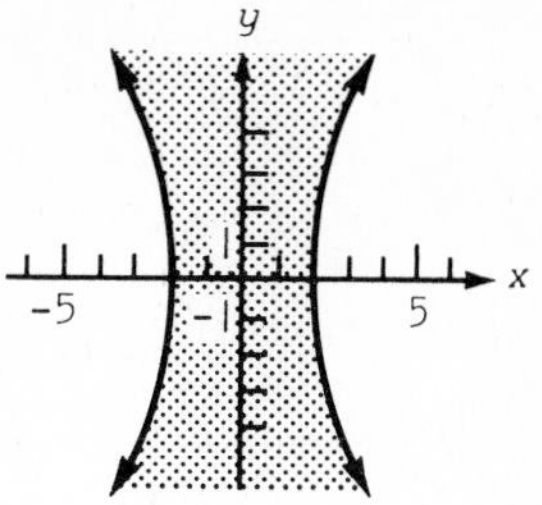

36. Problem 35 shows that $ax^2 - by^2 = c$ can be expressed as

$$y = \pm\sqrt{\frac{a}{b}}\,x\sqrt{1 - \frac{c}{ax^2}}\,.$$

$\frac{1}{x^2}$ approaches zero as x increases in value. Therefore, $\frac{c}{ax^2} = \frac{c}{a}\left(\frac{1}{x^2}\right)$ also approaches zero for large values of x, and thus $1 - \frac{c}{ax^2}$ approaches 1 for large values of x. Since $\sqrt{1 - \frac{c}{ax^2}}$ approaches $\sqrt{1} = 1$, $y = \pm\sqrt{\frac{a}{b}}\,x\sqrt{1 - \frac{c}{ax^2}}$ approaches $y = \pm\sqrt{\frac{a}{b}}\,x$. Hence, the hyperbola $ax^2 - by^2 = c$ approaches the lines $y = \pm\sqrt{\frac{a}{b}}\,x$. Therefore, the lines represented by $y = \pm\sqrt{\frac{a}{b}}\,x$ are asymptotes to the graph of $ax^2 - by^2 = c$.

EXERCISE 8.5

2. $T = ks$ 4. $t = \frac{k}{r}$ 6. $P = kRI^2$

8. Write an equation expressing the relationship between the variables: $y = kx$. Replace the variables by the given values and solve for k.

$$2 = k(5); \quad \text{therefore,} \quad k = 2/5.$$

10. Write an equation expressing the relationship between the variables: $r = \frac{k}{t^3}$. If $r = 8$ when $t = 10$, then

$$8 = \frac{k}{10^3}, \text{ from which } k = 8000.$$

12. $p = kqr$. If $p = 5$ when $q = 2$ and $r = 7$, then

$$5 = k(2)(7), \text{ from which } k = \frac{5}{14}.$$

14. $z = \frac{kx^3}{y^2}$. If $z = 2$ when $x = 2$ and $y = 4$, then

$$2 = \frac{k(2)^3}{4^2}, \text{ from which } k = 4.$$

16. $z = \frac{k(x + y)}{xy}$. If $z = 8$ when $x = 3$ and $y = 4$, then

$$8 = \frac{k(3 + 4)}{(3)(4)}, \text{ from which } k = \frac{96}{7}.$$

18. $r = \frac{ks}{t}$ (1). Let $r = 12$, $s = 8$, and $t = 2$. Then

$$12 = \frac{k(8)}{2}, \text{ from which } k = 3.$$

Substitute 3 for k in Equation (1): $r = \frac{3s}{t}$. Replace s and t by the second set of values, $s = 3$ and $t = 6$.

$$r = \frac{(3)(3)}{6}, \text{ from which } r = \frac{3}{2}.$$

20. Write an equation expressing the relationship between the variables: $d = kt^2$. From Problem 19, when $d = 16$, $t = 2$. Hence, $16 = k(2)^2$, from which $k = 4$. Thus, $d = 4t^2$. In this last equation let $t = 20$; then

$$d = 4(20)^2 = 1600.$$

Hence, the body will fall 1600 feet.

22. The equation expressing the relationship between the variables is $V = \frac{kT}{P}$ (1). Since $V = 20$ when $T = 300$ and $P = 30$,

$$20 = \frac{k(300)}{30}, \text{ from which } k = 2.$$

Replace k by 2 in Equation (1). $V = \frac{2T}{P}$ (2).

Since the temperature is raised to 360°K and the pressure is decreased to 20 pounds per square inch, we replace T by 360 and P by 20 in Equation (2), so that

$$V = \frac{2(360)}{20} = 36.$$

Hence, the volume will be 36 cubic feet.

24. $R = \frac{kl}{d^2}$ (1). Since $l = 50$, $d = 0.012$, when $R = 10$.

$$k = \frac{10(0.012)^2}{50} = 288 \times 10^{-7}.$$

Replace k by 288×10^{-7} in (1) and obtain $R = \frac{288 \times 10^{-7} l}{d^2}$

Now, for $l = 50$ and $d = 0.015$,

$$R = \frac{288 \times 10^{-7} \times 50}{(0.015)^2} = \frac{32}{5} = 6.4.$$

Hence, the resistance is 6.4 ohms.

26. $r = \frac{ks}{t}$. Solve for k: $k = \frac{rt}{s}$. Write a proportion for two different sets of conditions: $\frac{r_1 t_1}{s_1} = \frac{r_2 t_2}{s_2}$. Substitute the known values:

$$\frac{(12)(2)}{8} = \frac{r_2(6)}{3}, \text{ from which hence } r_2 = \frac{3}{2}.$$

28. From Problem 20, $d = kt^2$; $k = \frac{d}{t^2}$. Then

$$\frac{d_1}{{t_1}^2} = \frac{d_2}{{t_2}^2};\quad \frac{16}{2^2} = \frac{d_2}{(20)^2};\quad d_2 = 1600.$$

Hence, the body will fall 1600 feet.

30. From Problem 22, $V = \frac{kT}{P}$; $k = \frac{VP}{T}$. Then

$$\frac{V_1 P_1}{T_1} = \frac{V_2 P_2}{T_2};\quad \frac{(20)(30)}{300} = \frac{V_2(20)}{360};\quad V_2 = 36.$$

Hence, the volume will be 36 cubic feet.

32. From Problem 24, $R = \frac{kl}{d^2}$; $k = \frac{Rd^2}{l}$. Then

$$\frac{R_1 d_1{}^2}{l_1} = \frac{R_2 d_2{}^2}{l_2}; \quad \frac{10(0.012)^2}{50} = \frac{R_2(0.015)^2}{50};$$

$$R_2 = \frac{10 \times 12^2}{15^2} = \frac{32}{5} = 6.4.$$

Hence, the resistance is 6.4 ohms.

34. Let A_1 and A_2 be the areas of two circles whose radii are r_1 and r_2, respectively. Since $A = \pi r^2$,

$$A_1 = \pi r_1{}^2 \quad \text{and} \quad A_2 = \pi r_2{}^2.$$

Therefore, $\frac{A_1}{r_1{}^2} = \pi$ and $\frac{A_2}{r_2{}^2} = \pi$. Hence, $\frac{A_1}{r_1{}^2} = \frac{A_2}{r_2{}^2}$.

Multiplying both sides of the last equation by $\frac{r_1{}^2}{A_2}$ yields

$$\left(\frac{r_1{}^2}{A_2}\right)\frac{A_1}{r_1{}^2} = \left(\frac{r_1{}^2}{A_2}\right)\frac{A_2}{r_2{}^2}; \quad \frac{A_1}{A_2} = \frac{r_1{}^2}{r_2{}^2}.$$

36. Let f = frequency, t = tension, and l = length. Then $f = \frac{k\sqrt{t}}{l}$ (1). In (1), replace t by $4t$ and l by $2l$ and obtain

$$f = \frac{k\sqrt{4t}}{2l} = \frac{k \cdot 2\sqrt{t}}{2l} = \frac{k\sqrt{t}}{l}. \quad (2)$$

Comparing f in (1) with f in (2), we see that there is no change in f.

38. $y = \frac{k}{x}$.

For $k = 1$, $y = \frac{1}{x}$, $x > 0$.

For $k = 2$, $y = \frac{2}{x}$, $x > 0$.

As k increases, the graph falls more slowly to the right and is shifted upward.

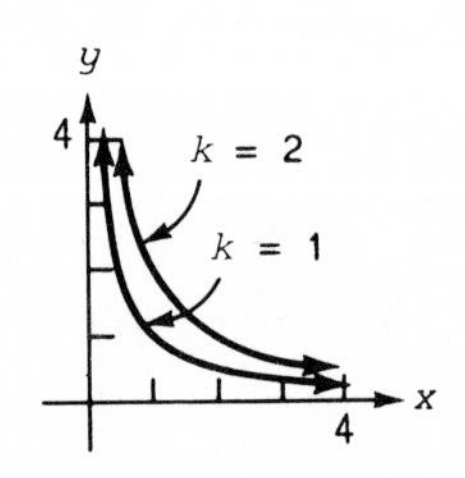

EXERCISE 8.6

2. Interchanging the components in each ordered pair:

$$f^{-1} = \{(1,-5),\ (2,5)\}.$$

f^{-1} is a function because the ordered pairs do not have the same first component.

4. Interchanging the components in each ordered pair:

$$q^{-1} = \{(2,2),\ (3,3),\ (3,4)\}.$$

q^{-1} is not a function because it has ordered pairs with the same first component.

6. $g^{-1} = \{(0,-2),\ (0,0),\ (-2,4)\}$. g^{-1} is not a function because it has ordered pairs with the same first component.

8. a. $3y - 2x = 5$

   b.

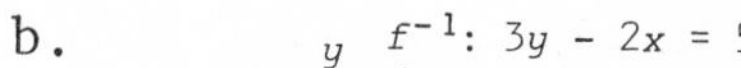

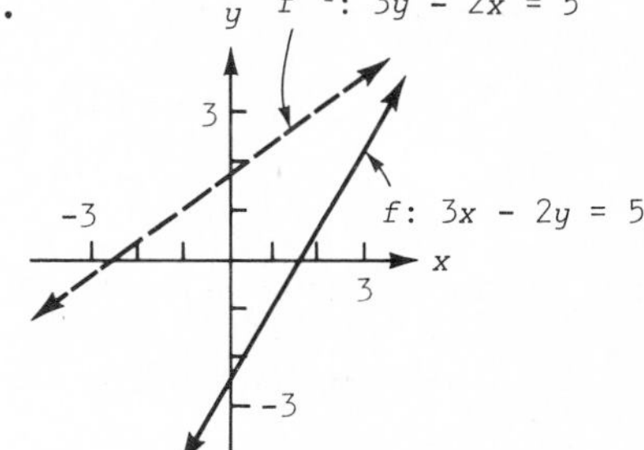

   c. A function because for each x value there is exactly one y value.

10. a. $x = y^2 - 4$

   b.

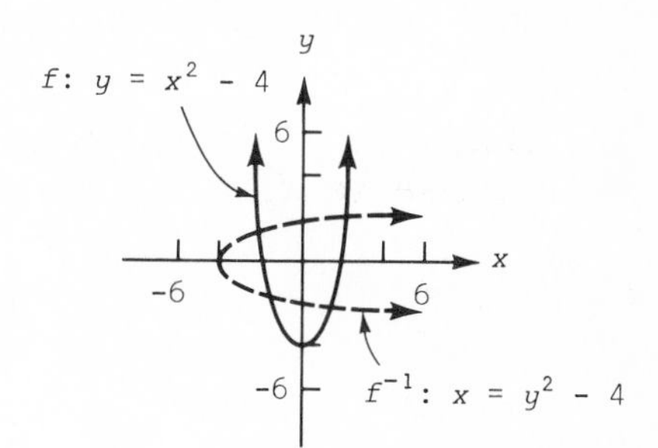

   c. Not a function because for each $x \geq -4$ there are two y values. For example, $(-3,1)$ and $(-3,-1)$ are members of f^{-1}.

12. a. $y^2 + x^2 = 4$

   b.

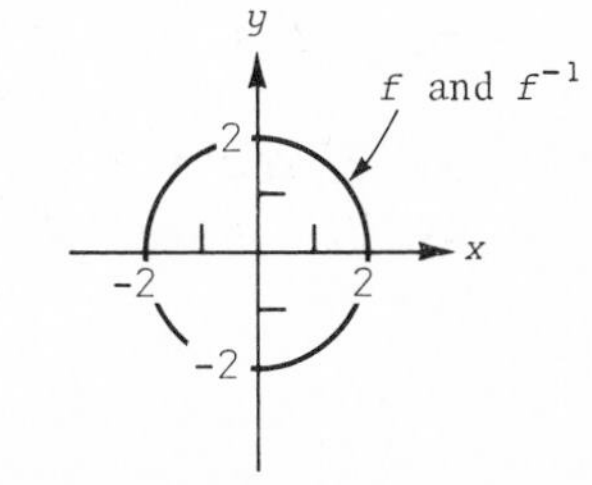

Note that the two graphs coincide.

   c. Not a function because for each $-2 \leq x \leq 2$ there are two y values. For example, $(1,\sqrt{3})$ and $(1,-\sqrt{3})$ are members of f^{-1}.

14. a. $9y^2 - x^2 = 36$

b.

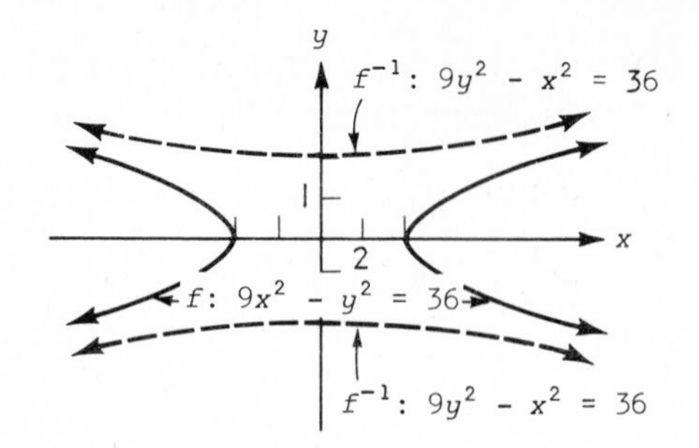

c. Not a function because for every real value of x there are two y values. For example, (0,2) and (0,-2) are members of f^{-1}.

16. a. $x = -\sqrt{y^2 - 4}$

b. To graph $y = -\sqrt{x^2 - 4}$, note that if both members are squared and the terms are rearranged, the equation can be written as $x^2 - y^2 = 4$, which we recognize as a hyperbola. We graph the part of the hyperbola that is on or below the x-axis because $y = -\sqrt{x^2 - 4}$. $x = -\sqrt{y^2 - 4}$ is similarly graphed.

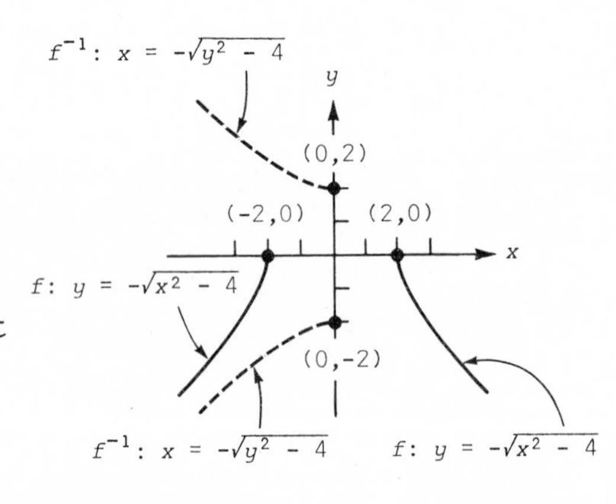

c. Not a function because for each $x < 0$ there are two y values. For example, $(-\sqrt{5},3)$ and $(-\sqrt{5},-3)$ are members of f^{-1}

18. a. $x = |y| + 1$

b.

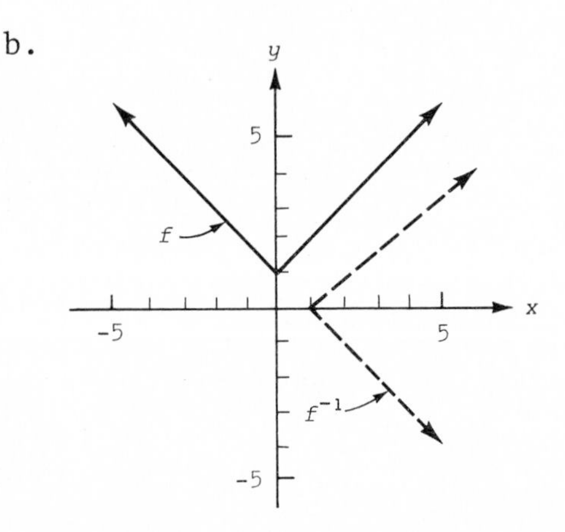

c. Not a function because for every x value there is more than one y value. For example, when $x = 2$, $y = \pm 1$.

20. $f: y = -x$; hence, $f(x) = -x$.
$f^{-1}: x = -y$ or $y = -x$; hence, $f^{-1}(x) = -x$.

$$f[f^{-1}(x)] = f[-x] = -(-x) = x$$
$$f^{-1}[f(x)] = f^{-1}[-x] = -(-x) = x$$

Thus, $f[f^{-1}(x)] = f^{-1}[f(x)] = x$.

22. $f: 2y = x - 4,\ y = \frac{1}{2}(x - 4)$; hence, $f(x) = \frac{1}{2}(x - 4)$.
$f^{-1}: y - 2x = 4,\ y = 2x + 4$; hence, $f^{-1}(x) = 2x + 4$.

$$f[f^{-1}(x)] = \frac{1}{2}[(2x + 4) - 4] = x$$
$$f^{-1}[f(x)] = 2[\frac{1}{2}(x - 4)] + 4 = x$$

Thus, $f[f^{-1}(x)] = f^{-1}[f(x)] = x$.

24. $f: 4y = -3x + 12,\ y = \frac{1}{4}(-3x + 12)$; hence, $f(x) = \frac{1}{4}(-3x + 12)$.
$f^{-1}: 3y + 4x = 12,\ y = \frac{1}{3}(-4x + 12)$; hence,
$f^{-1}(x) = \frac{1}{3}(-4x + 12)$.

$$f[f^{-1}(x)] = \frac{1}{4}[-3 \cdot \frac{1}{3}(-4x + 12) + 12]$$
$$= \frac{1}{4}[-(-4x + 12) + 12] = x$$

$$f^{-1}[f(x)] = \frac{1}{3}[-4 \cdot \frac{1}{4}(-3x + 12) + 12]$$
$$= \frac{1}{3}[-(-3x + 12) + 12] = x.$$

Thus, $f[f^{-1}(x)] = f^{-1}[f(x)] = x$.

26. A nonconstant linear function will have an oblique line as its graph. Hence, any horizontal line will intersect the graph in at most one point. Therefore, the nonconstant linear function has a function for an inverse.

28. a. Because $y = \sqrt{4 - x^2}$, $y \geqslant 0$.
Because $x = \pm\sqrt{4 - y^2}$ and $4 - y^2 \geqslant 0$, $-2 \leqslant y < 2$. Therefore, the range is given by

$$\{y \mid y \geqslant 0\} \cap \{y \mid -2 \leqslant y \leqslant 2\} = \{y \mid 0 \leqslant y \leqslant 2\}.$$

b. f^{-1} is defined by $x = \sqrt{4 - y^2}$. Hence, the domain is $x \geqslant 0$.
Because $y = \pm\sqrt{4 - x^2}$ and $4 - x^2 \geqslant 0$, $-2 \leqslant x \leqslant 2$. Therefore, the range is given by

$$\{x \mid x \geqslant 0\} \cap \{x \mid -2 \leqslant x \leqslant 2\} = \{x \mid 0 \leqslant x \leqslant 2\}.$$

c. f^{-1} is not a function because for x values in the domain there are two y values. For example, the ordered pairs $(\sqrt{3},1)$ and $(\sqrt{3},-1)$ are both members of f^{-1}.

9

EXPONENTIAL AND LOGARITHMIC FUNCTIONS

EXERCISE 9.1

2. For $x = \frac{-1}{2}$, $y = 4^{(-1/2)}$
$= \frac{1}{4^{1/2}}$
$= \frac{1}{2}$
$x = 0$, $y = 4^{(0)}$
$= 1$;
$x = \frac{1}{2}$, $y = 4^{(1/2)}$
$= 2$
The ordered pairs are
$\left(\frac{-1}{2},\frac{1}{2}\right)$, $(0,1)$, $\left(\frac{1}{2},2\right)$

4. For $x = -2$, $y = 5^{(-2)}$
$= \frac{1}{5^2} = \frac{1}{25}$;
$x = 0$, $y = 5^{(0)} = 1$;
$x = 2$, $y = 5^{(2)} = 25$.
The ordered pairs are
$\left(-2,\frac{1}{25}\right)$, $(0,1)$, $(2,25)$.

6. Observe that $\left(\frac{1}{3}\right)^x = \frac{1}{3^x}$
$= 3^{-x}$.
For $x = -3$, $y = 3^{-(-3)}$
$= 3^3 = 27$;
$x = 0$, $y = 3^{-(0)}$
$= 3^0 = 1$;
$x = 3$, $y = 3^{-(3)}$
$= 3^{-3} = \frac{1}{27}$.
The ordered pairs are
$(-3,27)$, $(0,1)$, $3,\frac{1}{27}$.

8. For $x = 0$, $y = 10^{(0)} = 1$;
$x = 1$, $y = 10^{(1)} = 10$;
$x = 2$, $y = 10^{(2)} = 100$.
The ordered pairs are
$(0,1)$, $(1,10)$, $(2,100)$.

10.

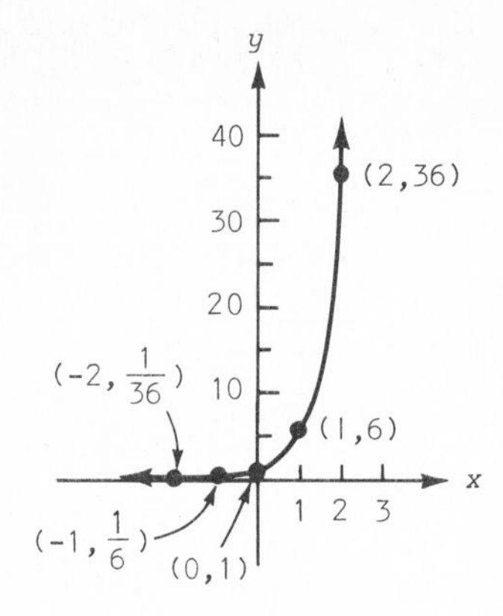

12.

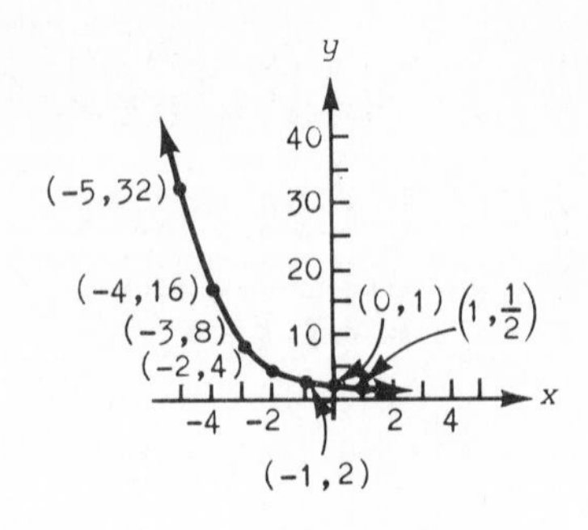

14. $y = 2^{2x} = (2^2)^x = 4^x$.

Hence, see the answer for Problem 9 of this section.

16. $y = \left(\frac{1}{3}\right)^x = \frac{1}{3^x} = 3^{-x}$

Hence, see the answer for Problem 13 of this section.

18. $y = \left(\frac{1}{10}\right)^x = 10^{-x}$

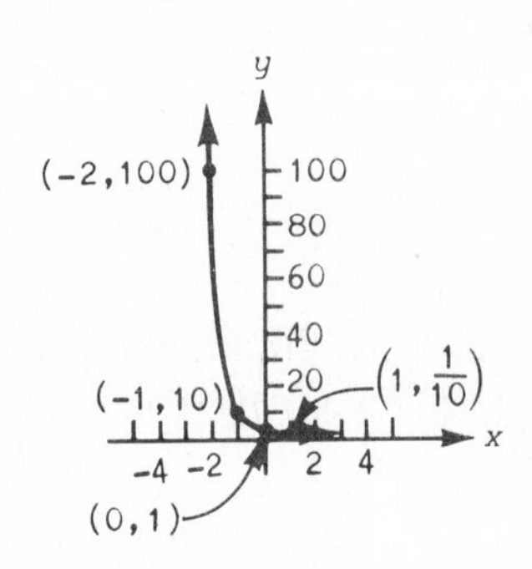

20. $y = \left(\frac{1}{3}\right)^{-x} = \frac{1}{3^{-x}} = 3^x$.

See example in text, p. 286

22. 5.01

24. 199.53

26. 0.06

28. 4.95

30. 1.54

32. 0.18

34. 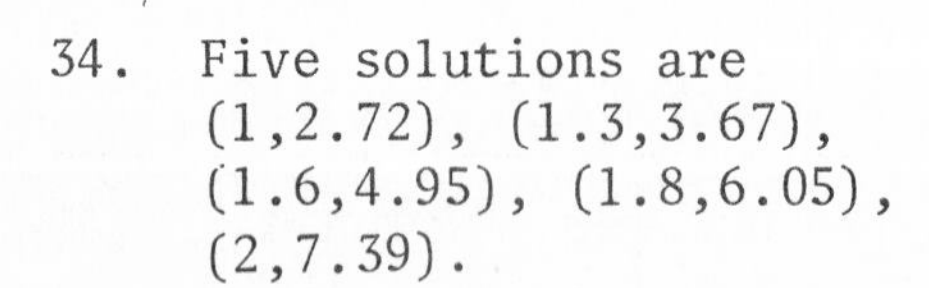
Five solutions are (1,2.72), (1.3,3.67), (1.6,4.95), (1.8,6.05), (2,7.39).

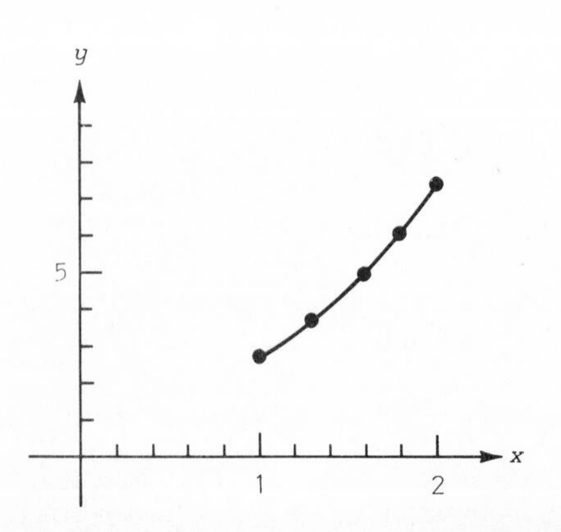

36. a. $P = 16{,}782{,}000\ e^{0.0083(10)}$
$= 16{,}782{,}000\ e^{0.083}$
$= 18{,}234{,}000$

b. In 1980:
$P = 16{,}782{,}000\ e^{0.0083(20)}$
$= 19{,}812{,}000$

In 1990:
$P = 16{,}782{,}000\ e^{0.0083(30)}$
$= 21{,}527{,}000$

In 2000:
$P = 16{,}782{,}000\ e^{0.0083(40)}$
$= 23{,}390{,}000$

38. $N = 6000\ e^{0.04(10)}$
$= 8951$

40. Graph $y_1 = 3^x$ and $y_2 = 4$ on the same axes. The solution is the x-coordinate of the point of intersection; $\left\{\frac{4}{3}\right\}$.

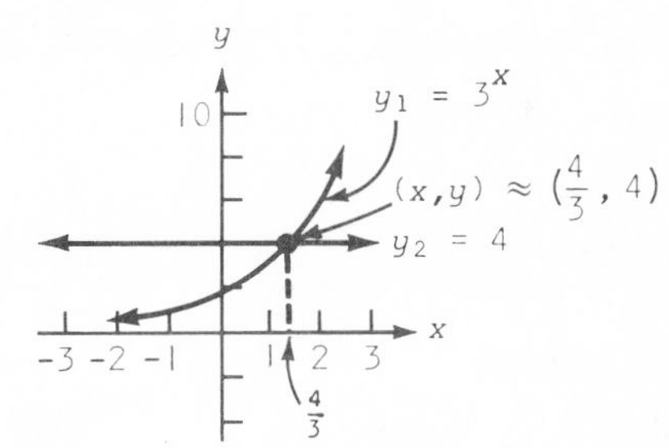

42. $a^{-x} = \frac{1}{a^x} = \left(\frac{1}{a}\right)^x$. Therefore, $y = a^{-x}$, a positive—or, equivalently, $y = \left(\frac{1}{a}\right)^x$, a positive—defines an increasing function, if the base $\frac{1}{a} > 1$. Solving this inequality for a, we have $a < 1$. Similarly, $y = a^{-x}$ defines a decreasing function if $\frac{1}{a} < 1$. Solving this inequality for a, we have $a > 1$.

EXERCISE 9.2

2. $\log_5 125 = 3$

4. $\log_8 64 = 2$

6. $\log_{1/3} \frac{1}{9} = 2$

8. $\log_{64} \frac{1}{2} = \frac{-1}{6}$

10. $\log_{10} 1 = 0$

12. $\log_{10} 0.01 = -2$

14. $5^2 = 25$

16. $16^2 = 256$

18. $\left(\frac{1}{2}\right)^{-3} = 8$

20. $10^0 = 1$

22. $10^{-4} = 0.0001$

24. $\log_2 32 = x$
$2^x = 32$
$x = 5$

26. $\log_5 \sqrt{5} = x$

$5^x = \sqrt{5} = 5^{1/2}$

$x = \frac{1}{2}$

28. $\log_3 \frac{1}{3} = x$

$3^x = \frac{1}{3} = 3^{-1}$

$x = -1$

30. $\log_3 3 = x$

$3^x = 3$

$x = 1$

32. $\log_{10} 10 = x$

$10^x = 10$

$x = 1$

34. $\log_{10} 1 = x$

$10^x = 1$

$x = 0$

36. $\log_{10} 0.01 = x$

$10^x = 0.01 = \frac{1}{10^2}$

$= 10^{-2}$

$x = -2$

38. $\log_5 125 = y$

$5^y = 125 = 5^3$

$y = 3$

40. $\log_b 625 = 4$

$b^4 = 625 = 5^4$

$b = 5$

42. $\log_{1/2} x = -5$

$x = \left(\frac{1}{2}\right)^{-5} = 2^5$

$= 32$

44. $\log_5 \frac{1}{5} = y$

$5^y = \frac{1}{5} = 5^{-1}$

$y = -1$

46. $\log_b 0.1 = -1$

$b^{-1} = 0.1$

$\frac{1}{b} = \frac{1}{10}$; $b = 10$

48. $\log_{10} x = -3$

$x = 10^{-3}$

$= 0.001$

50. $\log_5(\log_5 5)$

$= \log_5(1) = 0$

52. $\log_{10}[\log_2(\log_3 9)]$

$= \log_{10}[\log_2(2)]$

$= \log_{10}[1] = 0$

54. $\log_4[\log_2(\log_3 81)]$

$= \log_4[\log_2(4)]$

$= \log_4[2] = \frac{1}{2}$

56. Since $\log_a a^b = b$,

$\log_b(\log_a a^b) = \log_b(b) = 1.$

58. Since $\log_6 x$ is defined only if $x > 0$, it follows that $\log_6(x^2 - 4)$ is defined only if $x^2 - 4 > 0$. Hence, $x > 2$ or $x < -2$.

EXERCISE 9.3

2. $\log_b x + \log_b y$

4. $\log_b 4 + \log_b y + \log_b z$

6. $\log_b y - \log_b x$

8. $\log_b x - [\log_b(yz)] = \log_b x - [\log_b y + \log_b z]$
$= \log_b x - \log_b y - \log_b z$

10. $\frac{1}{3}\log_b x$

12. $\log_b y^{1/5} = \frac{1}{5}\log_b y$

14. $\log_b x^{3/2} = \frac{3}{2}\log_b x$

16. $\log_b x^{1/3} + \log_b z^2$
$= \frac{1}{3}\log_b x + 2\log_b z$

18. $\log_b xy^3 - \log_b z^{1/2} = \log_b x + \log_b y^3 - \log_b z^{1/2}$
$= \log_b x + 3\log_b y - \frac{1}{2}\log_b z$

20. $\log_{10}\left(\frac{x^2y}{z^3}\right)^{1/5} = \log_{10}\frac{x^{2/5}y^{1/5}}{z^{3/5}}$
$= \log_{10} x^{2/5} + \log_{10} y^{1/5} - \log_{10} z^{3/5}$
$= \frac{2}{5}\log_{10} x + \frac{1}{5}\log_{10} y - \frac{3}{5}\log_{10} z$

22. $\log_{10} 2y\left(\frac{x}{y}\right)^{1/3} = \log_{10} 2 + \log_{10} y + \log_{10}\left(\frac{x}{y}\right)^{1/3}$
$= \log_{10} 2 + \log_{10} y + \frac{1}{3}\left(\log_{10}\frac{x}{y}\right)$
$= \log_{10} 2 + \log_{10} y + \frac{1}{3}(\log_{10} x - \log_{10} y)$
$= \log_{10} 2 + \log_{10} y + \frac{1}{3}\log_{10} x - \frac{1}{3}\log_{10} y$

24. $\log_{10}\frac{2^{1/2}L^{1/2}}{R} = \frac{1}{2}\log_{10} 2 + \frac{1}{2}\log_{10} L - \log_{10} R$

26. $\log_{10} s(s - a)^{3/2}$
$= \log_{10} s + \frac{3}{2}\log_{10}(s - a)$

28. $\log_b\left(\frac{x}{y}\right)$

30. $\log_b x^{1/4} + \log_b y^{3/4} = \log_b x^{1/4}y^{3/4}$ or $\log_b \sqrt[4]{xy^3}$

32. $\frac{1}{3}(\log_b xy - \log_b z^2) = \frac{1}{3}\log_b\left(\frac{xy}{z^2}\right)$
$= \log_b\left(\frac{xy}{z^2}\right)^{1/3}$ or $\log_b \sqrt[3]{\frac{xy}{z^2}}$

34. $\frac{1}{2}(\log_{10} x - \log_{10} y^3 - \log_{10} z)$

$$= \frac{1}{2}(\log_{10} x - [\log_{10} y^3 + \log_{10} z])$$
$$= \frac{1}{2} \log_{10}\left(\frac{x}{y^3 z}\right) = \log_{10}\left(\frac{x}{y^3 z}\right)^{1/2} \quad \text{or} \quad \log_{10} \sqrt{\frac{x}{y^3 z}}$$

36. $-1 \cdot \log_b x = \log_b x^{-1}$ or $\log_b\left(\frac{1}{x}\right)$

38. $\log_{10}(x - 1) - \log_{10} 4 = \log_{10}\left(\frac{x - 1}{4}\right) = 2$

$$\frac{x - 1}{4} = 10^2$$
$$x - 1 = 400$$
$$x = 401; \quad \{401\}$$

40. $\log_{10}(x + 3) + \log_{10} x = \log_{10} x(x + 3) = 1$

$$x(x + 3) = 10^1$$
$$x^2 + 3x - 10 = 0$$
$$(x + 5)(x - 2) = 0$$
$$x = -5, 2.$$

We reject $x = -5$ since we must have $x > 0$ and $x + 3 > 0$. Hence, $x = 2$; $\{2\}$

42. $\log_{10}(x + 3) - \log_{10}(x - 1) = \log_{10}\left(\frac{x + 3}{x - 1}\right) = 1$

$$\frac{x + 3}{x - 1} = 10^1$$
$$x + 3 = 10(x - 1)$$
$$9x = 13,$$
$$x = \frac{13}{9}; \quad \left\{\frac{13}{9}\right\}$$

44.

$$\log_{10} 2p = \log_{10} k \cdot 10^{2t}$$
$$\log_{10} 2 + \log_{10} p = \log_{10} k + 2t \log_{10} 10$$
$$\log_{10} 2 + \log_{10} p = \log_{10} k + 2t(1)$$
$$\log_{10} 2 + \log_{10} p - \log_{10} k = 2t$$
$$\frac{\log_{10} 2 + \log_{10} p - \log_{10} k}{2} = t$$

46.
$$\log_{10} p = \log_{10} 200 + 30t \log_{10} 10$$
$$\log_{10} p = \log_{10}(100 \cdot 2) + 30t(1)$$
$$\log_{10} p = \log_{10} 100 + \log_{10} 2 + 30t$$
$$\log_{10} p = 2 + \log_{10} 2 + 30t$$
$$-2 + \log_{10} p - \log_{10} 2 = 30t$$
$$\frac{-2 + \log_{10} p - \log_{10} 2}{30} = t$$

48.
$$\log_{10} p = \log_{10} k + \frac{2t}{3} \log_{10} 10$$
$$\log_{10} p = \log_{10} k + \frac{2t}{3}(1)$$
$$\log_{10} p - \log_{10} k = \frac{2t}{3}$$
$$\frac{3}{2} \log_{10} p - \frac{3}{2} \log_{10} k = t$$

50. $\log_b 24 - \log_b 2 = \log_b \frac{24}{2}$
$= \log_b 12 = \log_b(3 \cdot 4) = \log_b 3 + \log_b 4$

52. $4 \log_b 3 - 2 \log_b 3 = 2 \log_b 3 = \log_b 3^2 = \log_b 9$

54. $\frac{1}{4} \log_b 8 + \frac{1}{4} \log_b 2 = \frac{1}{4}(\log_b 8 + \log_b 2)$
$= \frac{1}{4} \log_b(8 \cdot 2) = \log_b 16^{1/4} = \log_b 2$

56. Since $x_1 = b^{\log_b x_1}$,
$${x_1}^m = \left(b^{\log_b x_1}\right)^m$$
$${x_1}^m = b^{m \log_b x_1}.$$
By definition of a logarithm,
$\log_b {x_1}^m = m \log_b x_1$.

58. Let $x = 1000$ and $y = 100$.
$$\log_{10} \frac{1000}{100} = \log_{10} 10 = 1;$$
$$\frac{\log_{10} 1000}{\log_{10} 100} = \frac{3}{2}.$$
Since $1 \neq \frac{3}{2}$, $\log_{10} \frac{x}{y}$ is not equivalent to $\frac{\log_{10} x}{\log_{10} y}$.

EXERCISE 9.4

2. $\log_{10}(8.12 \times 10^0)$; 0

4. $\log_{10}(3.1 \times 10^1)$; 1

6. $\log_{10}(8.51 \times 10^{-3})$; -3 or 7 - 10

8. $\log_{10}(7.5231 \times 10^2)$; 2

10. $\log_{10}(4 \times 10^{-4})$; -4 or $6 - 10$

12. $\log_{10}(8.20 \times 10^{6})$; 6

14. $\log_{10}(8.91 \times 10^{2}) = 2.9499$

16. $\log_{10}(2.14 \times 10^{1}) = 1.3304$

18. $\log_{10}(2.19 \times 10^{2}) = 2.3404$

20. $\log_{10}(2.14 \times 10^{-3}) = -3 + 0.3304 = 7.3304 - 10$

22. $\log_{10}(4.13 \times 10^{-4}) = -4 + 0.6160 = 6.6160 - 10$

24. $-3 + 0.7316 = 7.7316 - 10$

26. 0.9128

28. 2.3096

30. 8.9614 - 10

32. 8.5729 - 10

34. $1.78 \times 10^{0} = 1.78$

36. $8.43 \times 10^{3} = 8430$

38. $9.38 \times 10^{-2} = 0.0938$

40. $4.56 \times 10^{2} = 456$

42. $1.02 \times 10^{2} = 102$

44. $6.54 \times 10^{-3} = 0.00654$

46. $6.44 \times 10^{0} = 6.44$

48. $1.83 \times 10^{2} = 183$

50. $7.49 \times 10^{-1} = 0.749$

52. 0.7577

54. 3.6896

56. $-1.1462 = -1.1462 + 10 - 10 = 8.8538 - 10$

58. $-3.2646 = -3.2646 + 10 - 10 = 6.7354 - 10$

60. $10^{0.4913} = 3.10$

62. $10^{3.7602} = 5760$

64. $10^{-2.0473} = 0.00897$

66. $10^{-3.8023} = 0.000158$

68. $-1.2984 = -1.2984 + 10 - 10 = 8.7016 - 10$

Then $\text{antilog}_{10}(-1.2984) = \text{antilog}_{10}(8.7016 - 10) = 5.03 \times 10^{-2} = .0503$

70. $-3.0670 = -3.0670 + 10 - 10 = 6.9330 - 10$

Then $\text{antilog}_{10}(-3.0670) = \text{antilog}_{10}(6.9330 - 10) = 8.57 \times 10^{-4} = 0.000857$

72. a. Since $\log_2\left(\frac{1}{8}\right) = -3$ and $\log_2\left(\frac{1}{4}\right) = -2$, and since $\frac{1}{8} < \frac{1}{5} < \frac{1}{4}$, it follows that $\log_2\left(\frac{1}{8}\right) < \log_2 \frac{1}{5} < \log_2\left(\frac{1}{4}\right)$, or equivalently, $\log_2\left(\frac{1}{5}\right)$ is between -3 and -2.

b. Since $\log_3\left(\frac{1}{3}\right) = -1$ and $\log_3\left(\frac{1}{9}\right) = -2$, and since $\frac{1}{9} < \frac{1}{7} < \frac{1}{3}$, it follows that $\log_3\left(\frac{1}{9}\right) < \log_3\left(\frac{1}{7}\right) < \log_3\left(\frac{1}{3}\right)$, or equivalently, $\log_3\left(\frac{1}{7}\right)$ is between -2 and -1.

74. $\text{antilog}_{10}\ 2.5761 = 10^{2.5761}$. Therefore,
$\log_{10}(\text{antilog}_{10}\ 2.5761) = \log_{10}(10^{2.5761}) = 2.5761$.

EXERCISE 9.5

2. Let $N = (82.3)(6.12)$.
Then $\log_{10} N = \log_{10} 82.3 + \log_{10} 6.12$.

$$\begin{aligned} \log_{10} 82.3 &= 1.9154 \\ \log_{10} 6.12 &= \underline{0.7868} \\ \log_{10} N &= 2.7022 \end{aligned}$$

$N = \text{antilog}_{10}\ 2.7022 = 5.04 \times 10^2$
$N = 504$

4. Let $N = \dfrac{1.38}{2.52}$.
Then $\log_{10} N = \log_{10} 1.38 - \log_{10} 2.52$.

$$\begin{aligned} \log_{10} 1.38 &= 10.1399 - 10 \\ \log_{10} 2.52 &= \underline{\ \ 0.4014\qquad\ } \\ \log_{10} N &= 9.7385 - 10 \end{aligned}$$

$N = \text{antilog}_{10}\ 9.7385 - 10$
$ = 5.48 \times 10^{-1}$ or
$N = 0.548$

6. Let $N = \dfrac{0.00214}{3.17}$.
Then $\log_{10} N = \log_{10} 0.00214 - \log_{10} 3.17$.

$$\begin{aligned} \log_{10} 2.14 \times 10^{-3} &= 7.3304 - 10 \\ \log_{10} 3.17 &= \underline{0.5011\qquad\quad} \\ \log_{10} N &= 6.8293 - 10 \end{aligned}$$

$N = \text{antilog}_{10}\ 6.8293 - 10$
$ = 0.000675$

8. Let $N = (4.62)^3$.
Then $\log_{10} N = 3 \log_{10} 4.62 = 3(0.6646) = 1.9938$.

$N = \text{antilog}_{10}\ 1.9938 = 98.6$

10. Let $N = \sqrt[5]{75} = (75)^{1/5}$.

Then $\log_{10} N = \frac{1}{5} \log_{10} 75 = \frac{1}{5}(1.8751) = 0.3750$.

$N = \text{antilog}_{10}\ 0.3750 = 2.37$

12. Let $N = (0.0021)^6$.

$$\begin{aligned}\text{Then } \log_{10} N &= 6 \log_{10}(2.1 \times 10^{-3}) \\ &= 6(0.3222 - 3) \\ &= 1.9332 - 18.\end{aligned}$$

$N = \text{antilog}_{10}(1.9332 - 18) = 8.57 \times 10^{-17}$

14. Let $N = \sqrt[5]{0.0471} = (0.0471)^{1/5}$

$$\begin{aligned}\text{Then } \log_{10} N &= \frac{1}{5} \log_{10}(4.71 \times 10^{-2}) \\ &= \frac{1}{5}(8.6730 - 10) \\ &= 1.7346 - 2.\end{aligned}$$

$N = \text{antilog}_{10}(1.7346 - 2) = 0.543$

16. Let $N = \sqrt[4]{0.0018} = (0.0018)^{1/4}$.

$$\begin{aligned}\text{Then } \log_{10} N &= \frac{1}{4} \log_{10} 0.0018 \\ &= \frac{1}{4}(37.2553 - 40) \\ &= 9.3138 - 10.\end{aligned}$$

$N = \text{antilog}_{10}(9.3138 - 10) = 0.206$

Note: 37 - 40 was chosen as the form of the characteristic because 40 is divisible by 4.

18. Let $N = \dfrac{(0.421)^2(84.3)}{\sqrt{21.7}}$.

$$\log_{10} N = 2 \log_{10}(4.21 \times 10^{-1}) + \log_{10}(8.43 \times 10) - \frac{1}{2} \log_{10}(2.17 \times 10)$$

$$= 2(0.6243 - 1) + 1.9258 - \frac{1}{2}(1.3365) = 0.5062$$

$N = \text{antilog}_{10}\ 0.5062 = 3.21$

20. Let $N = \dfrac{(2.61)^2(4.32)}{\sqrt{7.83}}$.

$$\log_{10} N = 2 \log_{10} 2.61 + \log_{10} 4.32 - \frac{1}{2} \log_{10} 7.83$$

$$= 2(0.4166) + 0.6355 - \frac{1}{2}(0.8938) = 1.0218$$

$N = \text{antilog}_{10}\ 1.0218 = 10.5$

22. Let $N = \dfrac{(4.813)^2(20.14)}{3.612}$.

$\log_{10} N = 2 \log_{10} 4.813 + \log_{10}(2.014 \times 10) - \log_{10} 3.612$

$\log_{10} N = 2(0.6821) + 1.3032 - 0.5575 = 2.1099.$

$N = \text{antilog}_{10}\ 2.1099 = 129$

24. $N = \sqrt{\dfrac{(2.85)^3(0.97)}{(0.035)}} = \dfrac{(2.85)^{3/2}(0.97)^{1/2}}{(0.035)^{1/2}}$

$\log_{10} N = \frac{3}{2} \log_{10} 2.85 + \frac{1}{2} \log_{10}(9.7 \times 10^{-1}) - \frac{1}{2} \log_{10}(3.5 \times 10^{-2})$

$= \frac{3}{2}(0.4548) + \frac{1}{2}(1.9868 - 2) - \frac{1}{2}(0.5441 - 2)$

$= 1.4036$

$N = \text{antilog}_{10}\ 1.4036 = 25.3$

26. $N = \sqrt{\dfrac{(4.17)^3(68.1 - 4.7)}{(68.1 - 52.9)}} = \dfrac{(4.17)^{3/2}(63.4)^{1/2}}{(15.2)^{1/2}}$

$\log_{10} N = \frac{3}{2} \log_{10} 4.17 + \frac{1}{2} \log_{10} 63.4 - \frac{1}{2} \log_{10} 15.2$

$= \frac{3}{2}(0.6201) + \frac{1}{2}(1.8021) - \frac{1}{2}(1.1818) = 1.2403$

$N = \text{antilog}_{10}\ 1.2403 = 17.4$

28. $N = \dfrac{\sqrt{38.7}\,\sqrt[3]{491}}{\sqrt[4]{9.21}} = \dfrac{(38.7)^{1/2}(491)^{1/3}}{(9.21)^{1/4}}$

$\log_{10} N = \frac{1}{2} \log_{10} 38.7 + \frac{1}{3} \log_{10}(491) - \frac{1}{4} \log_{10} 9.21$

$= \frac{1}{2}(1.5877) + \frac{1}{3}(2.6911) - \frac{1}{4}(0.9643) = 1.4498$

$N = \text{antilog}_{10}\ 1.4498 = 28.2$

30. $\log_{10} 50^{1/2} = \frac{1}{2} \log_{10}(5 \cdot 10) = \frac{1}{2}(\log_{10} 5 + \log_{10} 10)$

$= \frac{1}{2}(0.6990 + 1) = 0.8495$

32. $\log_{10} 0.08 - \log_{10} 15 = \log_{10}(8 \times 10^{-2}) - \log_{10}(3 \cdot 5)$

$= \log_{10}(2^3 \times 10^{-2}) - \log_{10}(3 \cdot 5)$

$= 3 \log_{10} 2 + \log_{10} 10^{-2} - \log_{10} 3 - \log_{10} 5$

$= 3(0.3010) + (-2) - 0.4771 - 0.6990$

$= -2.2731$ or $7.7269 - 10$

EXERCISE 9.6

2. $(2.303) \log_{10} 8$
$= (2.303)(0.9031)$
$= 2.08$

4. $(2.303) \log_{10} 98$
$= (2.303)(1.9912)$
$= 4.59$

6. $(2.303) \log_{10} 107$
$= (2.303)(2.0294)$
$= 4.67$

8. $(2.303) \log_{10} 14$
$= (2.303)(1.1461)$
$= 2.64$

10. $(2.303) \log_{10} 5$
$= (2.303)(0.6990)$
$= 1.61$

12. $(2.303) \log_{10} 18$
$= (2.303)(1.2553)$
$= 2.89$

14. $\log_{10} 3^x = \log_{10} 4$
$x(\log_{10} 3) = \log_{10} 4$
$\left\{\dfrac{\log_{10} 4}{\log_{10} 3}\right\}$
In decimal form:
$x = \dfrac{0.6021}{0.4771} = 1.26.$

16. $\log_{10} 2^{x-1} = \log_{10} 9$
$(x - 1) \log_{10} 2 = \log_{10} 9$
$x - 1 = \dfrac{\log_{10} 9}{\log_{10} 2}$
$\left\{\dfrac{\log_{10} 9}{\log_{10} 2} + 1\right\}$
In decimal form:
$x = \dfrac{0.9542}{0.3010} + 1 = 4.17.$

18. $\log_{10} 3^{x^2} = \log_{10} 21$
$x^2 \log_{10} 3 = \log_{10} 21$
$x^2 = \dfrac{\log_{10} 21}{\log_{10} 3}$
$\left\{\pm\sqrt{\dfrac{\log_{10} 21}{\log_{10} 3}}\right\}$
In decimal form:
$x \approx \pm\sqrt{\dfrac{1.3222}{0.4771}} = 1.66.$

20. $\log_{10} 2.13^{-x} = \log_{10} 8.1$
$-x \log_{10} 2.13 = \log_{10} 8.1$
$-x = \dfrac{\log_{10} 8.1}{\log_{10} 2.13}$
$\left\{\dfrac{-\log_{10} 8.1}{\log_{10} 2.13}\right\}$
In decimal form:
$x \approx \dfrac{-0.9085}{0.3284} = -2.77.$

22. $\ln 6.2 = \ln e^{x}$
$\ln 6.2 = x \ln e$
Since $\ln e = 1$,
$x = \ln 6.2$.

In decimal form:

$x = (2.303) \log_{10} 6.2$
$= (2.303)(0.7924)$
$= 1.82$.

24. $\ln 53.1 = \ln e^{x-2}$
$\ln 53.1 = (x - 2) \ln e$
Since $\ln e = 1$,
$\ln 53.1 = x - 2$;
$x = 2 + \ln 53.1$.

In decimal form:

$x = 2 + (2.303) \log_{10} 53.1$
$= 2 + (2.303)(1.7251)$
$= 5.97$.

26. $\ln 4.3 = \ln e^{-x}$
$\ln 4.3 = -x \ln e$
Since $\ln e = 1$,
$-x = \ln 4.3$;
$x = -\ln 4.3$.

In decimal form:

$x = (2.303)(-\log_{10} 4.3)$
$= (2.303)(-0.6335)$
$= -1.46$.

28. $\log_{10} y = \log_{10} Cx^{-n}$
$\log_{10} y = \log_{10} C - n \log_{10} x$
$n \log_{10} x = \log_{10} C - \log_{10} y$
$$n = \frac{\log_{10} C - \log_{10} y}{\log_{10} x}$$

30. $y = k - ke^{-t}$
$ke^{-t} = k - y$
$$e^{-t} = \frac{k - y}{k}$$
$$\ln e^{-t} = \ln\left(\frac{k - y}{k}\right)$$
$-t(\ln e) = \ln(k - y) - \ln k$
Since $\ln e = 1$,
$-t = \ln(k - y) - \ln k$
$t = -\ln(k - y) + \ln k$.

32. $$e^{-t/3} = \frac{B - 2}{A + 3}$$
$$\ln e^{-t/3} = \ln \frac{B - 2}{A + 3}$$
$$-\frac{t}{3}(\ln e) = \ln(B - 2) - \ln(A + 3)$$
Since $\ln \ = 1$,
$$-\frac{t}{3} = \ln(B - 2) - \ln(A + 3)$$
$t = -3 \ln(B - 2) + 3 \ln(A + 3)$.

34. Let $y = \log_2 3$. Then,

$$\begin{aligned} 2^y &= 3 \\ \log_{10} 2^y &= \log_{10} 3 \\ y \log_{10} 2 &= \log_{10} 3 \\ y &= \frac{\log_{10} 3}{\log_{10} 2} = \frac{0.4771}{0.3010} \\ &= 1.59 \end{aligned}$$

EXERCISE 9.7 APPLICATIONS

2.

$$\begin{aligned} 2.19 &= 1\left(1 + \frac{.04}{1}\right)^{1 \cdot n} \\ \log_{10}(1 + 0.04)^n &= \log_{10} 2.19 \\ n \log_{10} 1.04 &= \log_{10} 2.19 \\ n &= \frac{\log_{10} 2.19}{\log_{10} 1.04} = \frac{0.3404}{0.0170} = 20.02 \end{aligned}$$

It would take 20 years (to the nearest year).

4.

$$\begin{aligned} 50.90 &= 40\left(1 + \frac{r}{4}\right)^{12} \\ \left(1 + \frac{r}{4}\right)^{12} &= \frac{50.9}{40} \approx 1.272 \\ \log_{10}\left(1 + \frac{r}{4}\right)^{12} &= \log_{10} 1.272 \\ 12 \log_{10}\left(1 + \frac{r}{4}\right) &= \log_{10} 1.272 \\ \log_{10}\left(1 + \frac{r}{4}\right) &= \frac{\log_{10} 1.272}{12} = 0.0087 \\ 1 + \frac{r}{4} &= \text{antilog}_{10}\ 0.0087 = 1.02 \\ \frac{r}{4} &= 0.02, \quad r = 0.08 \quad \text{or} \quad r = 8\% \end{aligned}$$

6. A's amount after the first 3 months, or $\frac{1}{4}$ year, was

$$A = 10{,}000\left(1 + \frac{0.04}{4}\right)^{4 \cdot 1/4} = 10{,}000(1.01) = \$10{,}100.$$

Hence, his interest was \$100 for the 3-month period. This interest is withdrawn, and A's principal is once again \$10,000 for the second 3-month period (and for all following quarterly periods, because the procedure is repeated). Since there are four 3-month periods per year, A's interest for the 20 years is

$$20 \text{ years} \cdot 4 \frac{\text{periods}}{\text{year}} \cdot \$100/\text{period} = \$8000.$$

On the other hand, B's amount after 20 years is

$$B = 10{,}000\left(1 + \frac{0.04}{4}\right)^{4\cdot 20} = 10{,}000(1.01)^{80}.$$

$$\log_{10} B = \log_{10}[10^4(1.01)^{80}] = \log_{10} 10^4 + 80 \log_{10} 1.01$$
$$= 4 + 80(0.0043) = 4.3440.$$
$$B = \text{antilog}_{10}\ 4.3440 = 2.21 \cdot 10^4 = \$22{,}100.$$

Hence, B's interest for the 20 years was \$22,100 - \$10,000 = \$12,100, or \$4100 more than the \$8000 A earned.

8. $$\text{pH} = \log_{10}[H^+]$$
$$= \log_{10}(6.3 \times 10^{-7})$$
$$= -(0.7993 - 7)$$
$$= -0.7993 + 7$$
$$= 6.2$$

10. $$\text{pH} = \log_{10} \frac{1}{[H^+]} = 7.2$$
$$\frac{1}{[H^+]} = \text{antilog}_{10}\ 7.2$$
$$= 1.58 \times 10^7$$
$$[H^+] = \frac{1}{1.58 \times 10^7}$$
$$= 0.63 \times 10^{-7}$$
$$= 6.3 \times 10^{-8}$$

12. $$P = 30(10)^{-0.09(5.50)}$$
$$P = 30(10)^{-0.495}$$
$$\log_{10} P = \log_{10} 30 - 0.495 \log_{10} 10$$
Since $\log_{10} 30 = 1.4771$ and $\log_{10} 10 = 1$, we have
$$\log_{10} P = 1.4771 - 0.495(1)$$
$$= 0.9821.$$

$$P = \text{antilog}_{10}\ 0.9821 = 9.60$$

14. Let $a = 0$ to determine the pressure at sea level.
$$P = 30(10)^{-0.09(0)} = 30(10)^0$$
$$= 30(1) = 30$$
Then let $P = \frac{1}{2}(30) = 15$.
$$15 = 30(10)^{-0.09a}$$
Dividing by 30, we have
$$0.5 = 10^{-0.09a}$$
$$\log_{10} 0.5 = -0.09a \log_{10} 10$$
$$\log_{10} 0.5 = -0.09a(1)$$
$$a = \frac{-\log_{10} 0.5}{0.09} = \frac{-(-0.3010)}{0.09}$$
$$= 3.34$$

The altitude is 3.34 miles.

16. Let $P_0 = 4{,}951{,}600$, $P = 6{,}789{,}400$, and $t = 10$.

$$
\begin{aligned}
6{,}789{,}000 &= 4{,}951{,}600\ e^{10r} \\
\ln 6{,}789{,}000 &= \ln 4{,}951{,}600 + 10r \ln e \\
(2.303)(6.8319) &= (2.303)(6.6946) + 10r(1) \\
15.7339 &= 15.4177 + 10r \\
0.316 &\approx 10r \\
r &= 0.0316 = 3.16\%
\end{aligned}
$$

18. Let $y_0 = 40$ and $y = 12$.

$$
\begin{aligned}
12 &= 40e^{-0.4t} \\
\ln 12 &= \ln 40 - 0.4t \ln e \\
(2.303)(1.0792) &= (2.303)(1.6021) - 0.4t(1) \\
2.4854 &= 3.6896 - 0.4t \\
0.4t &= 1.2042 \\
t &= 3.01
\end{aligned}
$$

20. Let $I = 800$.

$$
\begin{aligned}
800 &= 1000e^{-0.1t} \\
\frac{800}{1000} &= e^{-0.1t} \\
e^{-0.1t} &= 0.8 \\
-0.1t \ln e &= \ln 0.8 \\
-0.1t(1) &= (2.303)(9.9031 - 10) \\
-0.1t &= (2.303)(-0.0969) \\
-0.1t &= -0.2232 \\
t &= 2.2
\end{aligned}
$$

22. Let the lengths of the three sides be $a = 2.314$, $b = 4.217$, and $c = 5.618$.

$$s = \frac{1}{2}(2.314 + 4.217 + 5.618) = \frac{1}{2}(12.149) = 6.075$$

$$s - a = 3.761; \quad s - b = 1.858; \quad s - c = 0.457$$

Then $A = \sqrt{(6.075)(3.761)(1.858)(0.457)}$, and

$$\log_{10} A = \frac{1}{2}[\log_{10} 6.075 + \log_{10} 3.761 + \log_{10} 1.858 + \log_{10} 0.457].$$

$$
\begin{aligned}
\log_{10} A &= \frac{1}{2}[0.7839 + 0.5752 + 0.2695 + 0.6599 - 1] \\
&= \frac{1}{2}[1.2885] = 0.6442
\end{aligned}
$$

$A = \text{antilog}_{10}\ 0.6442 = 4.41$ square inches

10

SYSTEMS OF EQUATIONS

EXERCISE 10.1

For these problems there are methods of solution other than those shown.

2. Add the two equations:

   $3x = 9; \quad x = 3.$

   Substitute 3 for x in the second equation:

   $3 + 3y = 3; \quad 3y = 0; \quad y = 0.$

   The solution set is $\{(3,0)\}$.

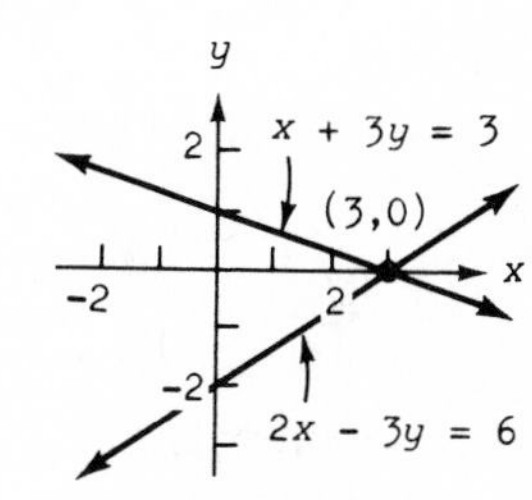

4. Add 2 times the first equation to the second:

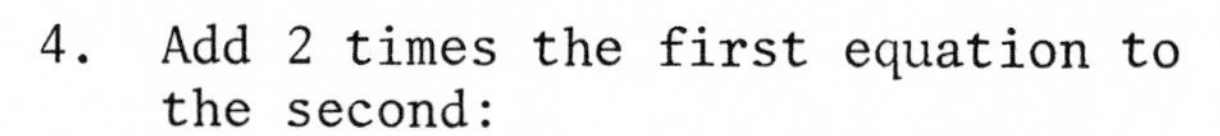

$$\begin{array}{l} 4x - 2y = 14 \\ \underline{3x + 2y = 14} \\ 7x \qquad = 28; \quad x = 4. \end{array}$$

   Substitute 4 for x in the first equation:

   $2(4) - y = 7; \quad 1 = y.$

   The solution set is $\{(4,1)\}$.

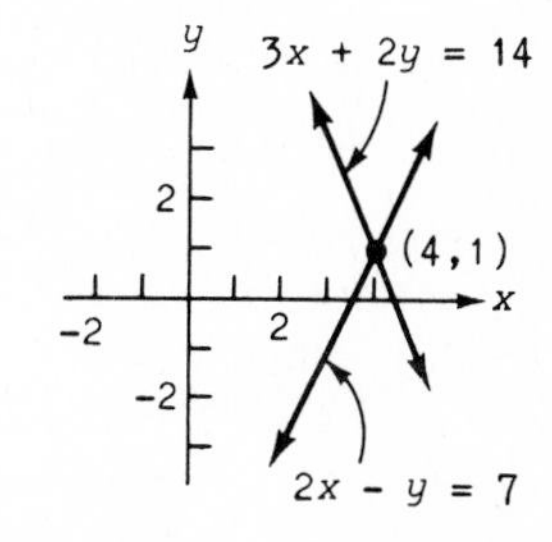

6. Add -2 times the second equation to the first:

$$\begin{array}{r} x + 4y = -14 \\ -6x - 4y = 4 \\ \hline -5x \quad = -10 \end{array}; \quad x = 2$$

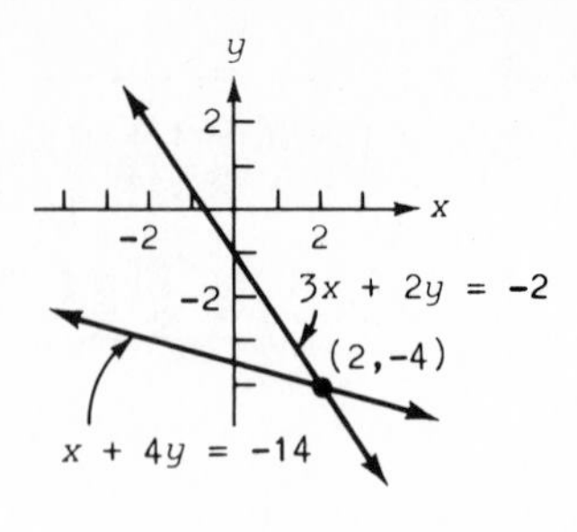

Substitute 2 for x in the first equation:

$2 + 4y = -14; \quad 4y = -16;$
$y = -4.$

The solution set is $\{(2,-4)\}$.

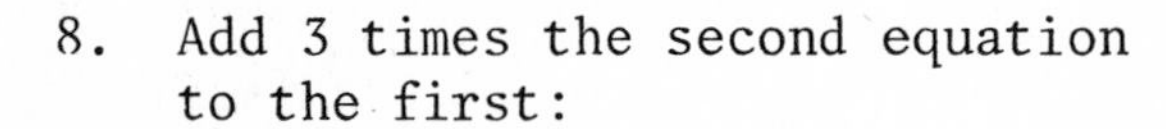

8. Add 3 times the second equation to the first:

$$\begin{array}{r} 2x - 3y = 8 \\ 3x + 3y = -3 \\ \hline 5x \quad = 5 \end{array}; \quad x = 1.$$

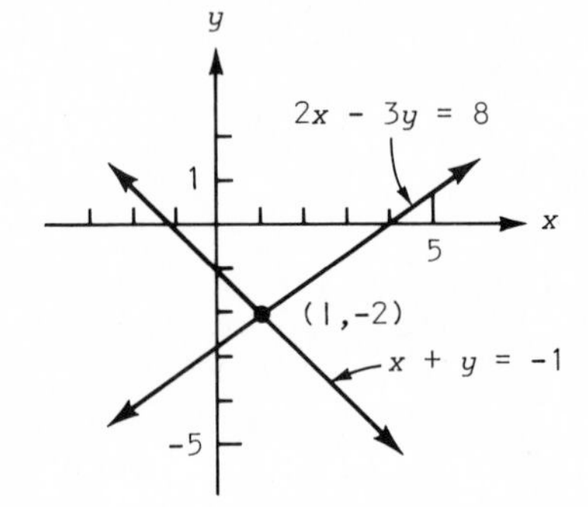

Substitute 1 for x in the second equation:

$1 + y = -1; \quad y = -2.$

The solution set is $\{(1,-2)\}$.

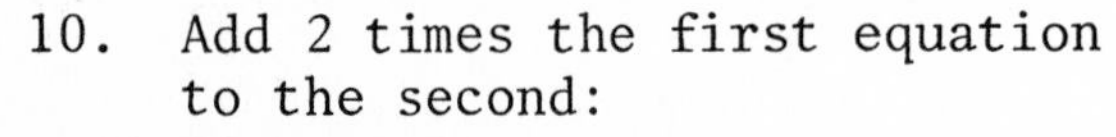

10. Add 2 times the first equation to the second:

$$\begin{array}{r} 6a - 6b = -6 \\ -6a + 2b = 14 \\ \hline -4b = 8 \end{array}; \quad b = -2.$$

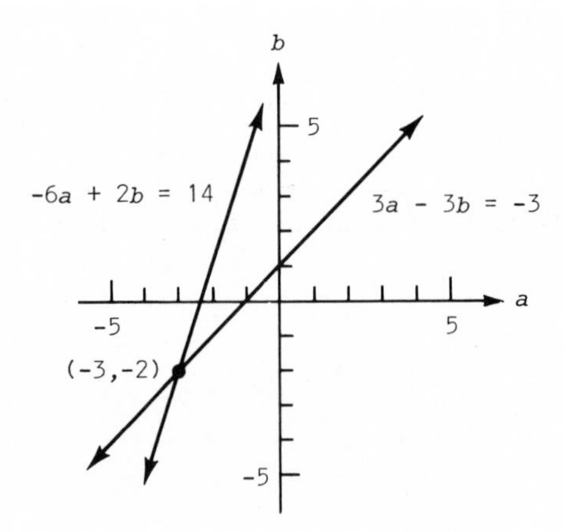

Substitute -2 for b in the first equation:

$3a - 3(-2) = -3; \quad a = -3.$

The solution set is $\{(-3,-2)\}$.

12. Add -3 times the second equation to the first:

$$\begin{array}{r} 3x - 5y = -1 \\ -3x - 6y = -54 \\ \hline -11y = -55 \end{array}; \quad y = 5.$$

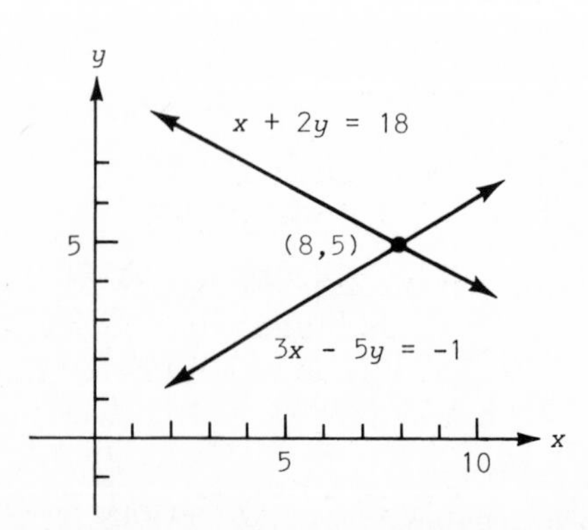

Substitute 5 for y in the second equation:

$x + 2(5) = 18; \quad x = 8.$

The solution set is $\{(8,5)\}$.

14. Substitute -3 for x in the first equation:

$2(-3) - y = 0; \quad -6 = y.$

The solution set is $\{(-3,-6)\}$.

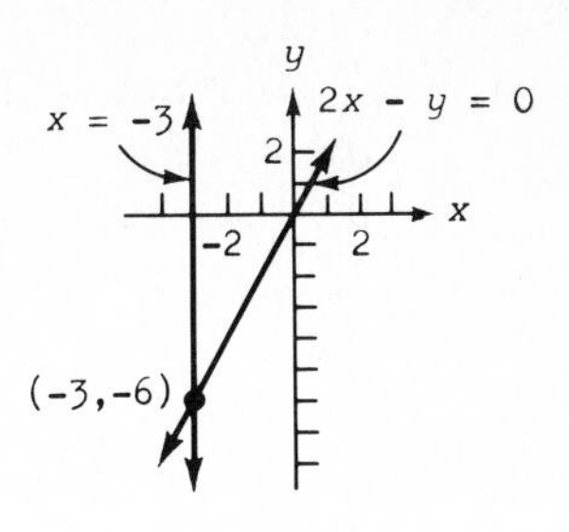

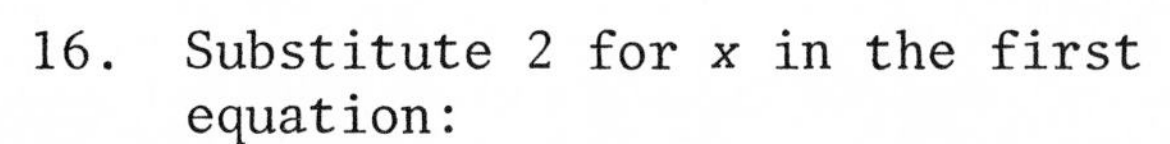
16. Substitute 2 for x in the first equation:

$2 + 2y = 6; \quad 2y = 4; \quad y = 2.$

The solution set is $\{(2,2)\}$.

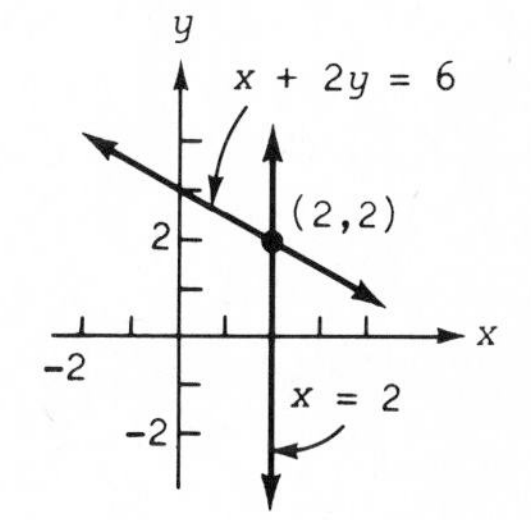

18. $\frac{2}{3}x - y = 4 \quad$ (A)

$x - \frac{3}{4}y = 6 \quad$ (B)

Multiply Equation (A) by 3 and Equation (B) by 4:

$2x - 3y = 12 \quad$ (A')
$4x - 3y = 24 \quad$ (B').

Add (-1) times Equation (A') to Equation (B'):

$2x = 12; \quad x = 6.$

Substitute 6 for x in Equation (A'):

$2(6) - 3y = 12;$
$12 - 12 = 3y;$
$0 = 3y; \quad 0 = y.$

The solution set is $\{(6,0)\}$.

20. $\frac{1}{3}x - \frac{2}{3}y = 2 \quad$ (A)

$x - 2y = 6 \quad$ (B)

Multiply Equation (A) by 3:

$x - 2y = 6 \quad$ (A')
$x - 2y = 6 \quad$ (B').

Since these equations are the same, any ordered pairs that satisfy one equation will satisfy the other; they are dependent.

The solution set is $\{(x,y) | x - 2y = 6\}$.

22. $3x - 2y = 6 \quad$ (A)
$6x - 4y = 8 \quad$ (B)

Multiply Equation (A) by -2 and add the resulting equation to Equation (B):

$0x + 0y = -4.$

Since this equation is not true for any values

24. $6x + 2y = 1 \quad$ (A)
$12x + 4y = 2 \quad$ (B)

Add -2 times Equation (A) to Equation (B):

$0x + 0y = 0.$

Since this equation is true for all values of x

of x and y, the solution set is $\emptyset$. The equations are inconsistent.

and y, the solution set has infinitely many solutions. The equations are dependent. The solution set is $\{(x,y) \mid 6x + 2y = 1\}$.

26. Let x represent the larger number and y the smaller.

$$x - y = 14 \quad (A)$$
$$x = 2y + 1 \quad \text{or} \quad x - 2y = 1 \quad (B)$$

Add (-1) times Equation (B) to Equation (A): $y = 13$.
From Equation (B): $x = 2(13) + 1 = 27$.

The numbers are 13 and 27.

28. Let x represent one integer and y the next.

$$y = x + 1 \quad \text{or} \quad x - y = -1 \quad (A)$$
$$\frac{1}{2}x + \frac{1}{5}y = 17 \quad \text{or} \quad 5x + 2y = 170 \quad (B)$$

Add 2 times Equation (A) to Equation (B):

$7x = 168$; $x = 24$.

From Equation (A): $y = 25$.

The integers are 24 and 25.

30. Let x and y be the number of votes cast for winner and loser, respectively:

$$x + y = 7179 \quad (A)$$
$$x - 6 = (y + 6) - 1 \quad \text{or} \quad x - y = 11 \quad (B)$$

Add Equation (A) to Equation (B): $2x = 7190$; $x = 3595$.

From Equation (A): $x + 3595 = 7179$; $x = 3584$.

The votes cast for the winner were 3595; for the loser, 3584.

32. Let x represent the amount invested at 4%; y the amount at 6%.

$$x + y = 1200 \quad (A)$$
$$0.04x = 0.06y + 3 \quad \text{or} \quad 4x - 6y = 300 \quad (B)$$

Multiply Equation (A) by 6 and add the resulting equation to Equation (B): $10x = 7500$; $x = 750$.

From Equation (A): $750 + y = 1200$; $y = 450$.

The amount invested at 4% was \$750; the amount at 6% was \$450.

34. Let s and d represent the ages in 1970 of the son and daughter, respectively.

$$s = 3d; \quad s - 3d = 0 \quad (A)$$

They received their funds 5 years after the father died in 1970; therefore,

$$s + 5 = 2(d + 5); \quad s - 2d = 5 \quad (B).$$

Add -1 times Equation (A) to Equation (B): $d = 5$.

From Equation (A): $s = 3(5) = 15$.

The son was 15 and the daughter was 5.

36. Substitute u for $\frac{1}{x}$ and v for $\frac{1}{y}$

$$u + 2v = \frac{-11}{12} \quad \text{or} \quad 12u + 24v = -11 \quad (A)$$

$$u + v = \frac{-7}{12} \quad \text{or} \quad 12u + 12v = -7 \quad (B)$$

Multiply Equation (B) by -1 and add to Equation (A):

$12v = -4; \quad v = \frac{-1}{3}$.

Substitute $\frac{-1}{3}$ for v in Equation (B):

$12u + 12\left(\frac{-1}{3}\right) = -7; \quad u = \frac{-3}{12} = \frac{-1}{4}$

Thus, $u = \frac{1}{x} = \frac{-1}{4}; \quad x = -4$, and $v = \frac{1}{y} = \frac{-1}{3}; \quad y = -3$.

The solution set is $\{(-4,-3)\}$.

38. Substitute u for $\frac{1}{x}$ and v for $\frac{1}{y}$.

$$u + 2v = 11 \quad (A)$$
$$u - 2v = -1 \quad (B)$$

Add Equations (A) and (B): $2u = 10; \quad u = 5$.

Substitute 5 for u in Equation (A): $5 + 2v = 11; \; v = 3$.

Thus, $u = \frac{1}{x} = 5; \quad x = \frac{1}{5}$, and $v = \frac{1}{y} = 3; \quad y = \frac{1}{3}$.

The solution set is $\left\{\left(\frac{1}{5}, \frac{1}{3}\right)\right\}$.

40. $\frac{2}{3}\left(\frac{1}{x}\right) + \frac{3}{4}\left(\frac{1}{y}\right) = \frac{7}{12}$; $4\left(\frac{1}{x}\right) - \frac{3}{4}\left(\frac{1}{y}\right) = \frac{7}{4}$.

Substitute u for $\frac{1}{x}$ and v for $\frac{1}{y}$.

$$\frac{2}{3}u + \frac{3}{4}v = \frac{7}{12};\quad 8u + 9v = 7 \quad (A)$$
$$4u - \frac{3}{4}v = \frac{7}{4};\quad 16u - 3v = 7 \quad (B)$$

Multiply Equation (B) by 3 and add to Equation (A):

$56u = 28;\quad u = \frac{1}{2}$

Multiply Equation (A) by -2 and add to Equation (B):

$-21v = -7;\quad v = \frac{1}{3}$.

Thus, $u = \frac{1}{x} = \frac{1}{2}$; $x = 2$, and $v = \frac{1}{y} = \frac{1}{3}$; $y = 3$.

The solution set is $\{(2,3)\}$.

42. Substitute 1 for x and 2 for y in the two equations.

$$a + 2b = 4 \quad (A)$$
$$b - 2a = -3 \text{ or } -2a + b = -3 \quad (B)$$

Add 2 times Equation (A) to Equation (B): $5b = 5$; $b = 1$.

Substitute 1 for b in Equation (A): $a + 2(1) = 4$; $a = 2$.

Thus, the solution is $a = 2$, $b = 1$.

44. $ax - by = 2$ (A)
$bx - ay = 1$ (B)

Multiply Equation (A) by a and (B) by $-b$ and add:

$a^2x - b^2x = 2a - b$; $(a^2 - b^2)x = 2a - b$; $x = \frac{2a - b}{a^2 - b^2}$

Now, multiply Equation (A) by $-b$ and (B) by a and add:

$b^2y - a^2y = -2b + a$; $(b^2 - a^2)y = -2b + a$;

$y = \frac{a - 2b}{b^2 - a^2} = \frac{2b - a}{a^2 - b^2}$; $\left\{\left(\frac{2a - b}{a^2 - b^2}, \frac{2b - a}{a^2 - b^2}\right)\right\}$.

46. $3x + 2y = 8$ (A)
$x - 3y = -5$ (B)
$-2x + y = 0$ (C)

Solve the system whose equations are (A) and (B). The unique solution is $\left(\frac{14}{11},\frac{23}{11}\right)$. Substitute $\frac{14}{11}$ for x and $\frac{23}{11}$ for y in Equation (C).

$$-2\left(\frac{14}{11}\right) + \frac{23}{11} \stackrel{?}{=} 0$$

Since this is a false statement, $\left(\frac{14}{11},\frac{23}{11}\right)$ is not a solution of Equation (C). Hence, the given system has no solution and the graphs of their equations have no point in common.

48. The results of Problem 47 are

$$x = \frac{b_2c_1 - b_1c_2}{a_1b_2 - a_2b_1}, \quad y = \frac{a_1c_2 - a_2c_1}{a_1b_2 - a_2b_1}.$$

Hence, taking $a_1 = 3$, $b_1 = 1$, $c_1 = 7$,
$a_2 = 2$, $b_2 = -5$, and $c_2 = -1$:

$$x = \frac{-5(7) - (1)(-1)}{3(-5) - (2)(1)} = 2; \quad y = \frac{3(-1) - (2)(7)}{-17} = 1. \quad \{(2,1)\}$$

EXERCISE 10.2

For these problems there are methods of solution other than those shown.

2.
$$\begin{aligned} x + y + z &= 1 \quad (1)\\ 2x - y + 3z &= 2 \quad (2)\\ 2x - y - z &= 2 \quad (3)\end{aligned}$$

Add Equations (1) and (2): $3x + 4z = 3$ (4).
Add Equations (1) and (3): $3x = 3$; or $x = 1$.

Substitute 1 for x in Equation (4): $3(1) + 4z = 3$; $z = 0$.
Substitute 1 for x and 0 for z in Equation (1):
$1 + y + 0 = 1$; $y = 0$.

The solution set is $\{(1,0,0)\}$.

4.
$$\begin{aligned} x - 2y + 4z &= -3 \quad (1)\\ 3x + y - 2z &= 12 \quad (2)\\ 2x + y - 3z &= 11 \quad (3)\end{aligned}$$

Add 2 times Equation (2) to Equation (1):
$7x = 21$; $x = 3$ (4).
Add 2 times Equation (3) to Equation (1):
$5x - 2z = 19$ (5).

Substitute 3 for x in Equation (5): $5(3) - 2z = 19$;
$-2z = 4$; $z = -2$.
Substitute 3 for x and -2 for z in Equation (2):

$3(3) + y - 2(-2) = 12$; $9 + y + 4 = 12$; $y = -1$.

The solution set is $\{(3,-1,-2)\}$.

6.
$$\begin{aligned} x + 5y - z &= 2 \quad (1)\\ 3x - 9y + 3z &= 6 \quad (2)\\ x - 3y + z &= 4 \quad (3)\end{aligned}$$

Add 3 times Equation (1) to Equation (2):

$6x + 6y = 12$; $x + y = 2$ (4).

Add Equation (1) to Equation (3):
$2x + 2y = 6$; $x + y = 3$ (5).

Since Equation (4) and Equation (5) are inconsistent, they have no points in common. Therefore, the given equations are inconsistent and the solution set is $\emptyset$.

8.
$$\begin{aligned} 2x - 3y + z &= 3 \quad (1)\\ x - y - 2z &= -1 \quad (2)\\ -x + 2y - 3z &= -4 \quad (3)\end{aligned}$$

Add 2 times Equation (1) to Equation (2): $5x - 7y = 5$ (4).
Add 3 times Equation (1) to Equation (3): $5x - 7y = 5$ (5).

Since Equation (4) is the same as Equation (5), their solutions are identical. The given equations are dependent.

10.
$$\begin{aligned} 5y - 8z &= -19 \quad (1)\\ 5x - 8z &= 6 \quad (2)\\ 3x - 2y &= 12 \quad (3)\end{aligned}$$

Add -1 times Equation (1) to Equation (2):

$5x - 5y = 25$; $x - y = 5$ (4).

Add -2 times Equation (4) to Equation (3): $x = 2$.

Substitute 2 for x in Equation (4): $2 - y = 5$; $-3 = y$.
Substitute -3 for y in Equation (1):

$5(-3) - 8z = -19$; $-8z = -4$; $z = \frac{1}{2}$.

The solution set is $\left\{\left(2,-3,\frac{1}{2}\right)\right\}$.

12.
$$\begin{aligned} x + 2y + \frac{1}{2}z &= 0 \quad (1)\\ x + \frac{3}{5}y - \frac{2}{5}z &= \frac{1}{5} \quad (2)\\ 4x - 7y - 7z &= 6 \quad (3)\end{aligned}$$

(continued)

Multiply Equation (1) by 2 and Equation (2) by 5.

$$\begin{aligned} 2x + 4y + z &= 0 \quad (4) \\ 5x + 3y - 2z &= 1 \quad (5) \\ 4x - 7y - 7z &= 6 \quad (3) \end{aligned}$$

Add 2 times Equation (4) to Equation (5):

$$9x + 11y = 1 \quad (6).$$

Add 7 times Equation (4) to Equation (3):

$$18x + 21y = 6 \quad (7).$$

Add -2 times Equation (6) to Equation (7): $-y = 4$; $y = -4$.

Substitute -4 for y in Equation (6):

$9x + 11(-4) = 1$; $9x = 45$; $x = 5$.

Substitute 5 for x and -4 for y in Equation (4):

$2(5) + 4(-4) + z = 0$; $z = 6$.

The solution set is $\{(5,-4,6)\}$.

14. Let $x, y,$ and z represent the three numbers.

$$\begin{aligned} &x + y + z = 2 && (1) \\ &x = y + z \quad \text{or} \quad x - y - z = 0 && (2) \\ &z = y - x \quad \text{or} \quad x - y + z = 0 && (3) \end{aligned}$$

Add Equation (2) to -1 times Equation (3): $-2z = 0$; $z = 0$.

Substitute 0 for z in Equations (1) and (2):

$$\begin{aligned} x + y &= 2 \quad (4); \\ x - y &= 0 \quad (5). \end{aligned}$$

Add Equations (4) and (5): $2x = 2$; $x = 1$.

Substitute 1 for x and 0 for z in Equation (1):

$1 + y + 0 = 2$; $y = 1$.

The numbers are 1, 1, and 0.

16.
$$\begin{aligned} &x + y + z = 155 && (1) \\ &x = y - 20 \quad \text{or} \quad x - y = -20 && (2) \\ &y = z + 5 \quad \text{or} \quad y - z = 5 && (3) \end{aligned}$$

Add Equation (2) to Equation (1): $2x + z = 135$ (4).
Add -1 times Equation (3) to Equation (1):

$$x + 2z = 150 \quad (5).$$

Add -2 times Equation (4) to Equation (5): $-3x = -120$;
$x = 40$.
Substitute 40 for x in Equation (2): $40 - y = -20$; $y = 60$.
Substitute 60 for y in Equation (3): $60 - z = 5$; $z = 55$.

The sides of the triangle are 40 inches, 60 inches, and 55 inches.

18. If the graph contains the points, then their coordinates must satisfy the equation.

Substitute -2 for x and 3 for y:

$(-2)^2 + 3^2 + a(-2) + b(3) + c = 0;\ -2a + 3b + c = -13 \quad (1).$

Substitute 1 for x and 6 for y:

$1^2 + 6^2 + a(1) + b(6) + c = 0;\ a + 6b + c = -37 \quad (2).$

Substitute 2 for x and 4 for y:

$2^2 + 4^2 + a(2) + b(4) + c = 0;\ 2a + 4b + c = -20 \quad (3).$

Add -1 times Equation (1) to Equation (2):

$$3a + 3b = -24, \quad \text{or} \quad a + b = -8 \quad (4).$$

Add -1 times Equation (2) to Equation (3):

$$a - 2b = 17 \quad (5).$$

Add 2 times Equation (4) to Equation (5): $3a = 1;\ a = \frac{1}{3}$.

Add -1 times Equation (5) to Equation (4): $3b = -25;$

$$b = \frac{-25}{3}.$$

Substitute $\frac{1}{3}$ for a and $\frac{-25}{3}$ for in Equation (1):

$$-2\left(\frac{1}{3}\right) + 3\left(\frac{-25}{3}\right) + c = -13;\quad \frac{-2}{3} - \frac{75}{3} + c = -13;$$

$$c = \frac{77}{3} - 13 = \frac{77}{3} - \frac{39}{3} = \frac{38}{3}.$$

The solution is $a = \frac{1}{3}$, $b = \frac{-25}{3}$, and $c = \frac{38}{3}$.

20. $x + y + 2z = 2 \quad (1)$
$2x - y + z = 3 \quad (2)$

Add the two equations: $3x + 3z = 5;\quad x = \frac{5 - 3z}{3}$.

Add -2 times Equation (1) to Equation (2):

$-3y - 3z = -1;\quad 3y + 3z = 1;\quad y = \frac{1 - 3z}{3}$

If we let z assume any value, we can find corresponding values for x and y.

For example, if $z = 1$, $x = \frac{5 - 3}{3} = \frac{2}{3}$, and $y = \frac{1 - 3}{3} = \frac{-2}{3}$.

And a solution is $\left(\frac{2}{3}, \frac{-2}{3}, 1\right)$.

Or, if $z = -2$, $x = \frac{5 + 6}{3} = \frac{11}{3}$, and $y = \frac{1 + 6}{3} = \frac{7}{3}$.

Thus, another solution is $\left(\frac{11}{3}, \frac{7}{3}, -2\right)$.

EXERCISE 10.3

2. $y = x^2 - 2x + 1$ (1)
$y + x = 3$ (2)

Solve Equation (2) explicitly for y: $y = 3 - x$.

Substitute $3 - x$ for y in Equation (1):

$3 - x = x^2 - 2x + 1$; $0 = x^2 - x - 2$; $0 = (x - 2)(x + 1)$;
$x = 2$; $x = -1$.

Substitute these values for x in Equation (2):

$y + 2 = 3$; $y = 1$.
$y + (-1) = 3$; $y = 4$.

The solution set is $\{(2,1), (-1,4)\}$.

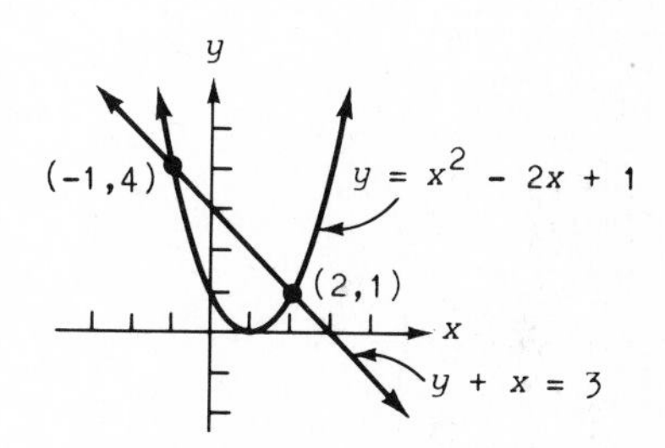

4. $x^2 + 2y^2 = 12$ (1)
$2x - y = 2$ (2)

Solve Equation (2) explicitly for y: $2x - 2 = y$.

Substitute $2x - 2$ for y in Equation (1):

$x^2 + 2(2x - 2)^2 = 12$; $x^2 + 2(4x^2 - 8x + 4) = 12$;
$9x^2 - 16x - 4 = 0$; $(9x + 2)(x - 2) = 0$;
$x = \frac{-2}{9}$; $x = 2$.

Substitute these values for x in Equation (2):

$2\left(\frac{-2}{9}\right) - y = 2$; $y = \frac{-22}{9}$;
$2(2) - y = 2$; $y = 2$.

The solution set is

$\left\{\left(\frac{-2}{9}, \frac{-22}{9}\right), (2,2)\right\}$.

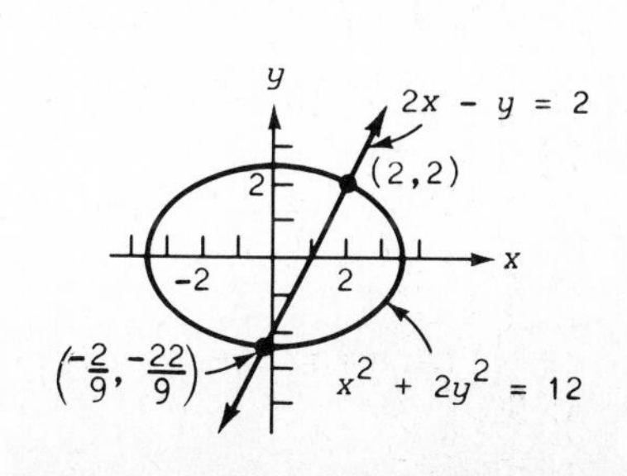

6. $2x - y = 9$ (1)
$xy = -4$ (2)

Solve Equation (1) for y: $2x - 9 = y$.
Substitute $2x - 9$ for y in Equation (2):

$x(2x - 9) = -4$; $2x^2 - 9x + 4 = 0$; $(2x - 1)(x - 4) = 0$;
$x = \frac{1}{2}$; $x = 4$.

Substitute these values for x in Equation (1):

$2\left(\frac{1}{2}\right) - y = 9$; $y = -8$;
$2(4) - y = 9$; $y = -1$;

The solution set is

$\left\{\left(\frac{1}{2},-8\right), (4,-1)\right\}$.

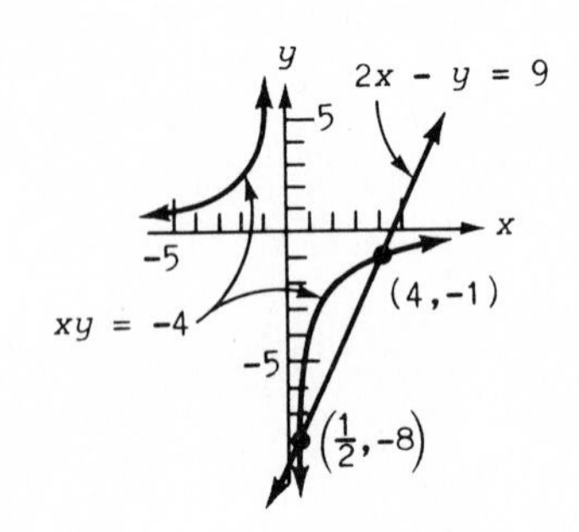

8. $x^2 - y^2 = 35$ (1)
$xy = 6$ (2)

Solve Equation (2) for y: $y = \frac{6}{x}$.

Substitute $\frac{6}{x}$ for y in Equation (1):

$x^2 - \left(\frac{6}{x}\right)^2 = 35$; $x^2 - \frac{36}{x^2} = 35$;
$x^4 - 35x^2 - 36 = 0$; $(x^2 - 36)(x^2 + 1) = 0$;
$x = \pm 6$; $x = \pm i$.

Substitute these values for x in Equation (2):

$6y = 6$ or $y = 1$; $-6y = 6$ or $y = -1$;
$iy = 6$; $y = \frac{6}{i}$ or $y = -6i$; $-iy = 6$; $y = \frac{6}{-i}$ or $y = 6i$.

The solution set is
$\{(6,1), (-6,-1), (i,-6i), (-i,6i)\}$.

Imaginary solutions do not show on the graph.

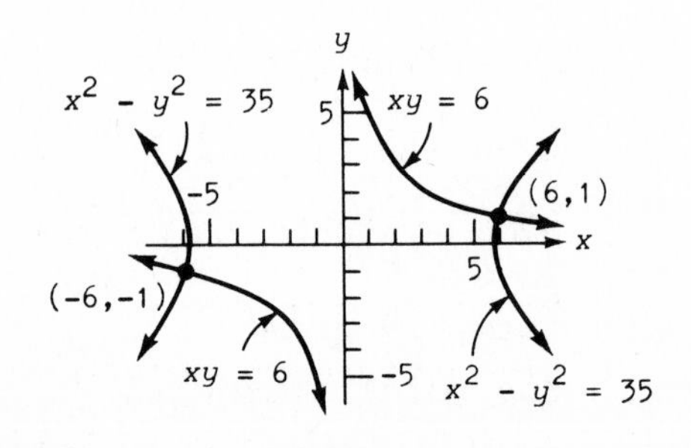

10. $2x^2 - 4y^2 = 12$ (1)
$x = 4$ (2)

Substitute 4 for x in Equation (1):

$2(4^2) - 4y^2 = 12;$
$32 - 4y^2 = 12;$
$5 = y^2; \quad y = \pm\sqrt{5}.$

The solution set is $\{(4,\sqrt{5}), (4,-\sqrt{5})\}$.

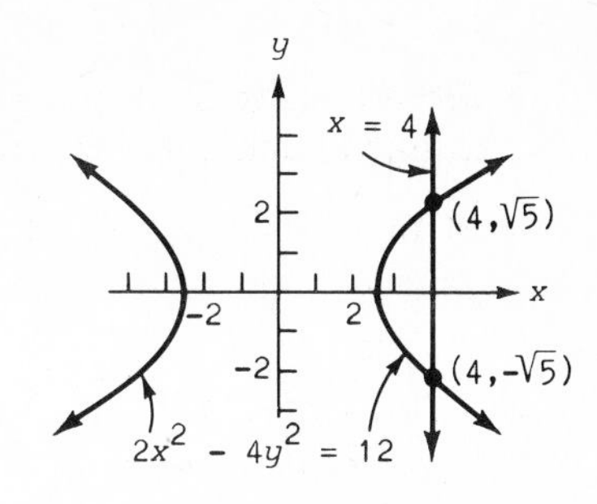

12. $x^2 + 9y^2 = 36$ (1)
$x - 2y = -8$ (2)

Solve Equation (2) for x: $x = 2y - 8$ (3).

Substitute $2y - 8$ for x in Equation (1):

$(2y - 8)^2 + 9y^2 = 36; \quad 13y^2 - 32y + 28 = 0.$

Use the quadratic formula:

$$y = \frac{32 \pm \sqrt{-432}}{26} = \frac{16 \pm 6i\sqrt{3}}{13}$$

Substitute these values for y in Equation (3):

$$x = 2\left(\frac{16 \pm 6i\sqrt{3}}{13}\right) - 8 = \frac{-72 \pm 12i\sqrt{3}}{13}.$$

The solution set is

$$\left\{\left(\frac{-72 + 12i\sqrt{3}}{13}, \frac{16 + 6i\sqrt{3}}{13}\right), \left(\frac{-72 - 12i\sqrt{3}}{13}, \frac{16 - 6i\sqrt{3}}{13}\right)\right\}.$$

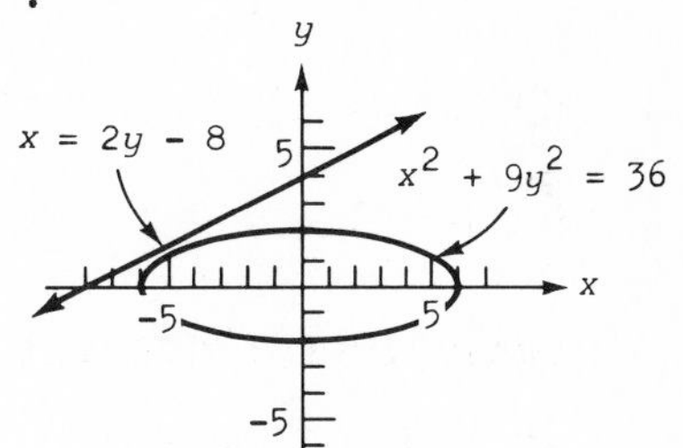

14. $x^2 - 2x + y^2 = 3$ (1)
$2x + y = 4$ (2)

Solve Equation (2) for y: $y = 4 - 2x$.

Substitute $4 - 2x$ for y in Equation (1):

$x^2 - 2x + (4 - 2x)^2 = 3; \quad 5x^2 - 18x + 13 = 0;$

$(5x - 13)(x - 1) = 0; \quad x = \frac{13}{5}; \quad x = 1.$

Substitute these values for x in Equation (2):

$2(1) + y = 4; \quad y = 2;$

$2\left(\frac{13}{5}\right) + y = 4; \quad y = \frac{-6}{5}.$

The solution set is $\left\{\left(\frac{13}{5}, \frac{-6}{5}\right), (1,2)\right\}$.

16. $2x^2 + xy + y^2 = 9$ (1)
$-x + 3y = 9$ (2)

Solve Equation (2) for x: $3y - 9 = x$.

Substitute $3y - 9$ for x in Equation (1):

$2(3y - 9)^2 + (3y - 9)y + y^2 = 9$; $22y^2 - 117y + 153 = 0$;
$(22y - 51)(y - 3) = 0$;
$y = \frac{51}{22}$; $y = 3$.

Substitute these values for y in Equation (2):

$-x + 3\left(\frac{51}{22}\right) = 9$; $x = \frac{-45}{22}$;
$-x + 3(3) = 9$; $x = 0$.

The solution set is $\left\{\left(\frac{-45}{22}, \frac{51}{22}\right), (0,3)\right\}$.

18. Let x and y represent the numbers.

$$x + y = 6 \quad (1)$$
$$xy = \frac{35}{4} \quad (2)$$

Solve Equation (1) for y: $y = 6 - x$.

Substitute $6 - x$ for y in Equation (2):

$x(6 - x) = \frac{35}{4}$; $24x - 4x^2 = 35$; $0 = (2x - 7)(2x - 5)$;
$x = \frac{7}{2}$; $x = \frac{5}{2}$.

Substitute these values in Equation (1):

$\frac{7}{2} + y = 6$; $y = \frac{5}{2}$.
$\frac{5}{2} + y = 6$; $y = \frac{7}{2}$.

The numbers are $\frac{5}{2}$ and $\frac{7}{2}$. The order is not important.

20. Let x represent the length and y represent the width of the rectangle.

$$xy = 216 \quad (1)$$
$$2x + 2y = 60 \quad (2)$$

Solve Equation (2) for y: $2y = 60 - 2x$; $y = 30 - x$.

Substitute $30 - x$ for y in Equation (1):

$x(30 - x) = 216; \quad 0 = x^2 - 30x + 216; \quad 0 = (x - 18)(x - 12).$
$x = 18; \quad x = 12.$

Substitute these values for x in Equation (2):

$2(18) + 2y = 60; \quad y = 12$
$2(12) + 2y = 60; \quad y = 18.$

Since the designation of length and width is unimportant, the rectangle is 18 feet by 12 feet.

22. Let x represent the amount invested and y represent the rate.

$$xy = 32 \quad (1)$$
$$(x + 200)(y - 0.005) = 35 \quad (2)$$

Solve Equation (1) for y: $y = \frac{32}{x}$.

Substitute $\frac{32}{x}$ for y in Equation (2):

$(x + 200)\left(\frac{32}{x} - 0.005\right) = 35; \quad 32 - 0.005x + \frac{6400}{x} - 1 = 35;$
$x^2 + 800x - 1{,}280{,}000 = 0; \quad (x - 800)(x + 1600) = 0;$
$x = 800; \quad x = -1600.$

Since the amount must be a positive number, $x = 800$.

Substitute 800 for x in Equation (1):

$800y = 32; \quad y = 0.04$ or 4%.

Thus, \$800 is invested at 4%.

24. $x^2 + y^2 = 25 \quad (1)$
$y = ax + b \quad (2)$

Substitute $ax + b$ for y in Equation (1):

$x^2 + (ax + b)^2 = 25; \quad x^2 + a^2x^2 + 2abx + b^2 = 25;$
$(a^2 + 1)x^2 + 2abx + (b^2 - 25) = 0.$

Using the coefficients $(a^2 + 1)$, $2ab$, and $(b^2 - 25)$, the discriminant of the quadratic formula is

$$(2ab)^2 - 4(a^2 + 1)(b^2 - 25) = 4a^2b^2 - 4a^2b^2 - 4b^2 + 100a^2 + 100$$
$$= 100a^2 - 4b^2 + 100.$$

a. To have two ordered pairs of real solutions, the discriminant must be positive. That is,

$$100a^2 - 4b^2 + 100 > 0; \quad b^2 < 25(a^2 + 1).$$

Hence, $-5\sqrt{a^2 + 1} < b < 5\sqrt{a^2 + 1}$.

b. To have one ordered pair as a solution, the discriminant must be zero. That is,

$100a^2 - 4b^2 + 100 = 0; \quad b^2 = 25(a^2 + 1); \quad b = \pm 5\sqrt{a^2 + 1}.$

c. To have no real solutions, the discriminant must be negative. That is,

$$100a^2 - 4b^2 + 100 < 0; \quad b^2 > 25(a^2 + 1).$$

Hence, $b > 5\sqrt{a^2 + 1}$ or $b < -5\sqrt{a^2 + 1}$.

26. The graphs have only two points in common, (2,2) and (-2,-2). Hence, these ordered pairs are the only ordered pairs of real numbers in the solution set of the given system of equations.

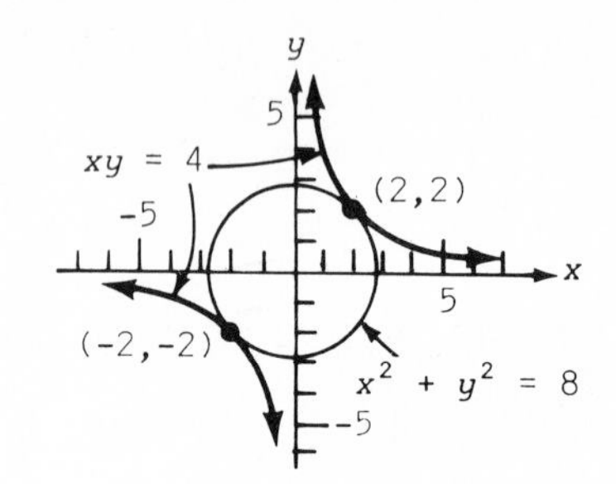

EXERCISE 10.4

2. $x^2 + 4y^2 = 52$ (1)
$x^2 + y^2 = 25$ (2)

Adding (-1) times Equation (2) to Equation (1):

$$3y^2 = 27; \quad y^2 = 9; \quad y = \pm 3.$$

Substituting these values for y in Equation (2):

$$x^2 + (3)^2 = 25; \quad x^2 = 16; \quad x = \pm 4;$$
$$x^2 + (-3)^2 = 25; \quad x^2 = 16; \quad x = \pm 4.$$

The solution set is {(4,3), (-4,3), (4,-3), (-4,-3)}.

4. $9x^2 + 16y^2 = 100$ (1)
$x^2 + y^2 = 8$ (2)

Adding -9 times Equation (2) to Equation (1):

$$7y^2 = 28; \quad y^2 = 4; \quad y = \pm 2.$$

Substituting these values for y in Equation (2);

$$x^2 + (2)^2 = 8; \quad x^2 = 4; \quad x = \pm 2;$$
$$x^2 + (-2)^2 = 8; \quad x^2 = 4; \quad x = \pm 2.$$

The solution set is {(2,2), (-2,2), (2,-2), (-2,-2)}.

6. $x^2 + 4y^2 = 25$ (1)
$4x^2 + y^2 = 25$ (2)

Adding -4 times Equation (2) to Equation (1):

$$-15x^2 = -75; \quad x^2 = 5; \quad x = \pm\sqrt{5}.$$

Substituting these values for x in Equation (2):

$$4(\sqrt{5})^2 + y^2 = 25; \quad y^2 = 5; \quad y = \pm\sqrt{5};$$
$$4(-\sqrt{5})^2 + y^2 = 25; \quad y^2 = 5; \quad y = \pm\sqrt{5}.$$

The solution set is $\{(\sqrt{5},\sqrt{5}), (\sqrt{5},-\sqrt{5}), (-\sqrt{5},\sqrt{5}), (-\sqrt{5},-\sqrt{5})\}$.

8. $4x^2 + 3y^2 = 12$ (1)
$x^2 + 3y^2 = 12$ (2)

Adding -1 times Equation (2) to Equation (1):

$$3x^2 = 0; \quad x = 0.$$

Substituting 0 for x in Equation (2):

$$0 + 3y^2 = 12; \quad y^2 = 4; \quad y = \pm 2.$$

The solution set is $\{(0,2), (0,-2)\}$.

10. $16y^2 + 5x^2 - 26 = 0$ (1)
$25y^2 - 4x^2 - 17 = 0$ (2)

Adding 4 times Equation (1) to 5 times Equation (2):

$$189y^2 - 189 = 0; \quad y^2 = 1; \quad y = \pm 1.$$

Substituting these values for y in Equation (1):

$$16(1)^2 + 5x^2 - 26 = 0; \quad x^2 = 2; \quad x = \pm\sqrt{2};$$
$$16(-1)^2 + 5x^2 - 26 = 0; \quad x^2 = 2; \quad x = \pm\sqrt{2}.$$

The solution set is $\{(\sqrt{2},1), (-\sqrt{2},1), (\sqrt{2},-1), (-\sqrt{2},-1)\}$.

12. $x^2 + 2xy - y^2 = 14$ (1)
$x^2 - y^2 = 8$ (2)

Adding -1 times Equation (2) to Equation (1):

$$2xy = 6; \quad xy = 3 \quad (3).$$

Solving this equation for y: $y = \dfrac{3}{x}$

Substituting $\dfrac{3}{x}$ for y in Equation (2):

$$x^2 - \left(\frac{3}{x}\right)^2 = 8; \quad x^4 - 8x^2 - 9 = 0; \quad (x^2 - 9)(x^2 + 1) = 0;$$
$$x^2 - 9 = 0; \quad x^2 + 1 = 0; \quad x = \pm 3; \quad x = \pm i.$$

Substituting these values for x in Equation (3):

$3y = 3$; $y = 1$; $-3y = 3$; $y = -1$;
$(i)y = 3$; $y = \frac{3}{i}$; $y = -3i$; $(-i)y = 3$; $y = \frac{3}{-i}$; $y = 3i$.

The solution set is $\{(3,1), (-3,-1), (i,-3i), (-i,3i)\}$.

14. $2x^2 + xy - 2y^2 = 16$ (1)
$x^2 + 2xy - y^2 = 17$ (2)

Adding -2 times Equation (2) to Equation (1):

$$-3xy = -18;$$
$$xy = 6 \quad (3).$$

Solving this equation for y: $y = \frac{6}{x}$.

Substituting $\frac{6}{x}$ for y in Equation (2):

$$x^2 + 2x\left(\frac{6}{x}\right) - \left(\frac{6}{x}\right)^2 = 17; \quad x^4 - 5x^2 - 36 = 0;$$
$$(x^2 - 9)(x^2 + 4) = 0;$$
$$x^2 - 9 = 0; \quad x^2 + 4 = 0; \quad x = \pm 3; \quad x = \pm 2i.$$

Substituting these values for x in Equation (3):

$3y = 6$; $y = 2$; $-3y = 6$; $y = -2$;
$2iy = 6$; $y = \frac{6}{2i}$; $y = -3i$; $-2iy = 6$; $y = \frac{6}{-2i}$; $y = 3i$.

The solution set is $\{(3,2), (-3,-2), (2i,-3i), (-2i,3i)\}$.

16. $x^2 - xy + y^2 = 21$ (1)
$x^2 + 2xy - 8y^2 = 0$ (2)

If $x^2 + 2xy - 8y^2 = (x + 4y)(x - 2y) = 0$, then

$$x + 4y = 0 \quad (3); \quad \text{or}$$
$$x - 2y = 0 \quad (4).$$

Solve Equation (3) for x: $x = -4y$.

Substitute in Equation (1):

$$(-4y)^2 - (-4y)y + y^2 = 21; \quad y^2 = 1; \quad y = \pm 1.$$

Substituting these values for y in Equation (3): $x = \pm 4$.

Two solutions are $(4,-1)$, $(-4,1)$.

Solve Equation (4) for x: $x = 2y$.

Substitute in Equation (1):

$$(2y)^2 - (2y)y + y^2 = 21; \quad y^2 = 7; \quad y = \pm\sqrt{7}.$$

Substituting these values for y in Equation (4): $x = \pm 2\sqrt{7}$.

The solution set is $\{(4,-1), (-4,1), (2\sqrt{7},\sqrt{7}), (-2\sqrt{7},-\sqrt{7})\}$.

18. $\dfrac{1}{x^2} + \dfrac{3}{y^2} = 7 \quad (1)$

$\dfrac{2}{x^2} - \dfrac{5}{y^2} = 3 \quad (2)$

Substituting u^2 for $\dfrac{1}{x^2}$ and v^2 for $\dfrac{1}{y^2}$:

$$u^2 + 3v^2 = 7 \quad (3);$$
$$2u^2 - 5v^2 = 3 \quad (4).$$

Adding -2 times Equation (3) to Equation (4):

$$-11v^2 = -11; \quad v^2 = 1.$$

Substituting this value of v^2 into Equation (3):

$$u^2 + 3(1) = 7; \quad u^2 = 4.$$

Substituting 4 for u^2 in $u^2 = \dfrac{1}{x^2}$ and 1 for v^2 in $v^2 = \dfrac{1}{y^2}$:

$$4 = \frac{1}{x^2}, \quad x = \frac{\pm 1}{2}; \quad 1 = \frac{1}{y^2}, \quad y = \pm 1.$$

The solution set is $\left\{\left(\frac{1}{2},1\right), \left(\frac{1}{2},-1\right), \left(\frac{-1}{2},1\right), \left(\frac{-1}{2},-1\right)\right\}$.

11

NATURAL NUMBER FUNCTIONS

EXERCISE 11.1

2. $s_1 = 2(1) - 3 = -1;$ $\quad s_2 = 2(2) - 3 = 1;$
   $s_3 = 2(3) - 3 = 3;$ $\quad s_4 = 2(4) - 3 = 5$

4. $s_1 = \dfrac{3}{1^2 + 1} = \dfrac{3}{2};$ $\quad s_2 = \dfrac{3}{(2)^2 + 1} = \dfrac{3}{5};$
   $s_3 = \dfrac{3}{(3)^2 + 1} = \dfrac{3}{10};$ $\quad s_4 = \dfrac{3}{(4)^2 + 1} = \dfrac{3}{17}$

6. $s_1 = \dfrac{1}{2(1) - 1} = 1;$ $\quad s_2 = \dfrac{2}{2(2) - 1} = \dfrac{2}{3};$
   $s_3 = \dfrac{3}{2(3) - 1} = \dfrac{3}{5};$ $\quad s_4 = \dfrac{4}{2(4) - 1} = \dfrac{4}{7}$

8. $s_1 = \dfrac{5}{1(1 + 1)} = \dfrac{5}{2};$ $\quad s_2 = \dfrac{5}{2(2 + 1)} = \dfrac{5}{6};$
   $s_3 = \dfrac{5}{3(3 + 1)} = \dfrac{5}{12};$ $\quad s_4 = \dfrac{5}{4(4 + 1)} = \dfrac{1}{4}$

10. $s_1 = (-1)^{1+1} = 1;$ $\quad s_2 = (-1)^{2+1} = -1;$
    $s_3 = (-1)^{3+1} = 1;$ $\quad s_4 = (-1)^{4+1} = -1$

12. $s_1 = (-1)^{1-1}3^{1+1} = 9;$ $\quad s_2 = (-1)^{2-1}3^{2+1} = -27;$
    $s_3 = (-1)^{3-1}3^{3+1} = 81;$ $\quad s_4 = (-1)^{4-1}3^{4+1} = -243$

14. $[3(1) - 2] + [3(2) - 2] + [3(3) - 2] = 1 + 4 + 7$

16. $(2^2 + 1) + (3^2 + 1) + (4^2 + 1) + (5^2 + 1) + (6^2 + 1)$
$= 5 + 10 + 17 + 26 + 37$

18. $\frac{2}{2}(2 + 1) + \frac{3}{2}(3 + 1) + \frac{4}{2}(4 + 1) + \frac{5}{2}(5 + 1) + \frac{6}{2}(6 + 1)$
$= 3 + 6 + 10 + 15 + 21$

20. $\frac{(-1)^{3+1}}{3 - 2} + \frac{(-1)^{4+1}}{4 - 2} + \frac{(-1)^{5+1}}{5 - 2} = 1 - \frac{1}{2} + \frac{1}{3}$

22. $1 + \frac{1}{2} + \frac{1}{3} + \cdots$

24. $\frac{0}{1 + 0} + \frac{1}{1 + 1} + \frac{2}{1 + 2} + \cdots = 0 + \frac{1}{2} + \frac{2}{3} + \cdots$

26. $\sum_{i=1}^{4} 2i$

28. $\sum_{i=1}^{5} x^{2i+1}$ or $\sum_{i=0}^{4} x^{2i+3}$

30. These are the cubes of the first five natural numbers:

$$\sum_{i=1}^{5} i^3.$$

32. Each denominator gives the number of the term, while each numerator is one greater than its denominator: $\sum_{i=1}^{\infty} \frac{i + 1}{i}$.

34. Each numerator is two more than the denominator:

$$s_\infty = \sum_{i=1}^{\infty} \frac{2i + 1}{2i - 1}.$$

36. The numerators are powers of 3. The demoninators are multiples of 2: $s_\infty = \sum_{i=1}^{\infty} \frac{3^{i-1}}{2i}$.

EXERCISE 11.2

In Problems 2-12, substitute into the equation $s_n = a + (n - 1)d$.

2. $d = -1 - (-6) = 5$. The sequence is

$$-6, -1, -1 + 5 = 4, 4 + 5 = 9, 9 + 5 = 14$$

$$s_n = -6 + (n - 1)5 = 5n - 11.$$

4. $d = -20 - (-10) = -10$. The sequence is

$$-10, -20, -30, -40, -50, \cdots$$

$$s_n = -10 + (n - 1)(-10) = -10n.$$

6. $d = (a + 5) - a = 5$. The sequence is

$$a, a + 5, a + 10, a + 15, a + 20, \cdots$$

$$s_n = a + (n - 1)5 = a - 5 + 5n.$$

8. $d = y - (y - 2b) = 2b$. The sequence is

$$y - 2b, y, y + 2b, y + 4b, y + 6b, \cdots$$

$$s_n = (y - 2b) + (n - 1)2b = y - 4b + 2bn.$$

10. $d = (a - 2b) - (a + 2b) = -4b$. The sequence is

$$a + 2b, a - 2b, a - 6b, a - 10b, a - 14b, \cdots$$

$$s_n = (a + 2b) + (n - 1)(-4b) = a + 6b - 4bn.$$

12. $d = 5a - 3a = 2a$. The sequence is

$$3a, 5a, 7a, 9a, 11a, \cdots$$

$$s_n = 3a + (n - 1)2a = a + 2an.$$

In Problems 14-18, first find the general term for each progression using the formula $s_n = a + (n - 1)d$, and then replace n by the appropriate natural number.

14. $d = -12 - (-3) = -9$; $a = -3$;

$$s_n = -3 + (n - 1)(-9) = 6 - 9n;$$
$$s_{10} = 6 - 9(10) = -84.$$

16. $d = (-2) - (-5) = 3; \quad a = -5;$

$$s_n = -5 + (n - 1)3 = 3n - 8;$$
$$s_{17} = 3(17) - 8 = 43.$$

18. $d = 2 - \frac{3}{4} = \frac{5}{4}; \quad a = \frac{3}{4};$

$$s_n = \frac{3}{4} + (n - 1)\frac{5}{4} = \frac{5n}{4} - \frac{1}{2};$$
$$s_{10} = \frac{5(10)}{4} - \frac{1}{2} = 12.$$

20. A diagram is helpful.

$n = 1 \quad 2 \quad 3 \quad 4 \quad 5 \quad 6 \cdots 12 \cdots 19 \quad 20$

$s_n = \_ \; \_ \; \_ \; \_ \; -16 \; \cdots \; ? \; \cdots \; \_ \; \underline{-46}$

Find d: $s_{16} = -46 = -16 + (16 - 1)d; \quad -46 = -16 + 15d;$
$d = -2.$

Find a: $s_5 = -16 = a + (5 - 1)(-2); \quad a = -8.$

$$s_n = -8 + (n - 1)(-2) = -6 - 2n;$$
$$s_{12} = -6 - 2(12) = -30.$$

Alternative solution:
Use $s_n = a + (n - 1)d$ twice; once with $s_5 = -16$ and again with $s_{20} = -46$ to obtain the system

$$-16 = a + (5 - 1)d$$
$$-46 = a + (20 - 1)d.$$

Solve the system to obtain $d = -2$ and $a = -8$. Then use $s_n = a + (n - 1)d$, with $n = 12$, $a = -8$, and $d = -2$:

$$s_{12} = -8 + (12 - 1)(-2) = -30.$$

22. $a = 7, \quad d = 3 - 7 = -4$

$$s_n = a + (n - 1)d = 7 + (n - 1)(-4) = 11 - 4n$$

Let $s_n = -81 = 11 - 4n; \quad n = 23;$

-81 is the 23rd term.

24. $\underline{10} \; \underline{?} \; \underline{?} \; \underline{?} \; \underline{?} \; \underline{65}$, with a difference d between each consecutive pair

There are 5 differences (d) between 10 and 65. Hence,

$$5d = 65 - 10; \quad d = 11$$

By four successive additions of 11, the required arithmetic means are 21, 32, 43, 54.

26. $\underline{-11}$ $\underline{\ ?\ }$ $\underline{\ 7\ }$

There are 2 differences (d) between -11 and 7. Hence,

$$2d = 7 - (-11); \quad d = 9.$$

Adding 9 to -11, the required mean is -2.

28. $\underline{-12}$ $\underline{\ ?\ }$ $\underline{\ ?\ }$ $\underline{\ ?\ }$ $\underline{\ ?\ }$ $\underline{\ ?\ }$ $\underline{\ ?\ }$ $\underline{23}$

d d d d d d d

There are 7 differences (d) between -12 and 23. Hence,

$$7d = 23 - (-12); \quad d = 5.$$

By six successive additions of 5, the required means are -7, -2, 3, 8, 13, 18.

30. $\displaystyle\sum_{i=1}^{21} (3i - 2) = 1 + 4 + 7 + \cdots; \quad d = 3, \quad a = 1, \quad n = 21.$

Using $s_n = \frac{n}{2}[2a + (n - 1)d]$,

$$s_{21} = \frac{21}{2}[2(1) + (21 - 1)3] = 651.$$

32. $\displaystyle\sum_{j=10}^{20} (2j - 3) = 17 + 19 + 21 + \cdots; \quad d = 2, \quad a = 17, \quad n = 20 - 9 = 11.$

Using $s_n = \frac{n}{2}[2a + (n - 1)d]$,

$$s_{11} = \frac{11}{2}[2(17) + (11 - 1)2] = 297.$$

34. $\displaystyle\sum_{k=1}^{100} k = 1 + 2 + 3 + \cdots; \quad d = 1, \quad a = 1, \quad n = 100.$

Using $s_n = \frac{n}{2}[2a + (n - 1)d]$,

$$s_{100} = \frac{100}{2}[2(1) + (100 - 1)1] = 5050.$$

36. The first term is $14 = 7 \cdot 2$; that is, the 2nd multiple of seven; and the last is $105 = 7 \cdot 15$, the 15th multiple of seven. Thus,

$$s_n = \sum_{i=2}^{15} 7i = 14 + 21 + 28 + \cdots + 105; \quad d = 7, \quad a = 14, \quad n = 15 - 1 = 14.$$

$$s_{14} = \frac{14}{2}[2(14) + (14 - 1)7] = 833.$$

38. The sum of the number of bricks in each row is an arithmetic series with the sum known to be 256.

$$\sum_{j=1}^{n} (2j - 1) = 1 + 3 + \cdots + (2n - 1) = 256; \quad a = 1, \quad d = 2.$$

$$s_n = 256 = \frac{n}{2}[2(1) + (n - 1)2]$$

From which, $n^2 = 256$; $n = 16$. That is, there are 16 rows; hence, the third row from the bottom is the 14th row. Using $s_n = a + (n - 1)d$ with $a = 1$, $d = 2$, $n = 14$:

$$s_{14} = 1 + (14 - 1)2 = 27.$$

There are 27 bricks in the third row from the bottom.

40.

a: the first number
$a + d$: the second number
$a + 2d$: the third number

$$a + (a + d) + (a + 2d) = 21 \quad \text{or} \quad 3a + 3d = 21 \qquad (1)$$
$$a(a + d)(a + 2d) = 231 \qquad (2)$$

Solve Equation (1) for d: $d = 7 - a$.

Substitute $7 - a$ for d in Equation (2):

$$\begin{aligned} a(a + (7 - a))(a + 2(7 - a)) &= 231 \\ 7a(-a + 14) &= 231 \\ -7a^2 + 98a - 231 &= 0 \\ a^2 - 14a + 33 &= 0 \\ (a - 11)(a - 3) &= 0 \end{aligned}$$

Therefore, $a = 11$ and $d = 7 - a = -4$, or $a = 3$ and $d = 7 - a = 4$. The required numbers are 11, 7, 3, or 3, 7, 11.

42. $[p + q] + [2p + q] + [3p + q] + [4p + q] = 28;$

$$10p + 4q = 28 \qquad (1)$$

$[2p + q] + [3p + q] + [4p + q] + [5p + q] = 44;$

$$14p + 4q = 44 \qquad (2)$$

Subtract Equation (1) from Equation (2): $4p = 16$; $p = 4$.

Substitute 4 in Equation (1): $10(4) + 4q = 28$; $q = -3$.
Hence, $p = 4$, $q = -3$.

44. $s_n = 2 + 4 + 6 + \cdots + 2n$; $a = 2$, $d = 2$, $n = n$.

$$s_n = \frac{n}{2}[2a + (n - 1)d] = \frac{n}{2}[2(2) + (n - 1)2]$$
$$= \frac{n}{2}[2n + 2] = n^2 + n.$$

EXERCISE 11.3

2. $r = \frac{8}{4} = 2$; $a = 4$. The next three terms are

$16 \cdot 2 = 32$, $32 \cdot 2 = 64$, $64 \cdot 2 = 128$, or 32, 64, 128.

$$s_n = 4(2)^{n-1} = 2^2(2)^{n-1}$$
$$= (2)^{n-1+2} = 2^{n+1}$$

4. $r = \frac{3}{6} = \frac{1}{2}$; $a = 6$. The next three terms are

$\frac{3}{2} \cdot \frac{1}{2} = \frac{3}{4}$, $\frac{3}{4} \cdot \frac{1}{2} = \frac{3}{8}$, $\frac{3}{8} \cdot \frac{1}{2} = \frac{3}{16}$ or $\frac{3}{4}$, $\frac{3}{8}$, $\frac{3}{16}$.

$$s_n = 6\left(\frac{1}{2}\right)^{n-1}$$

6. $r = \frac{-3}{2} \div \frac{1}{2} = -3$; $a = \frac{1}{2}$. The next three terms are

$$\frac{-27}{2}, \quad \frac{81}{2}, \quad \frac{-243}{2}.$$

$$s_n = \frac{1}{2}(-3)^{n-1}$$

8. $r = \frac{a}{bc} \div \frac{a}{b} = \frac{1}{c}$; the first term is $\frac{a}{b}$. The next three terms are

$$\frac{a}{bc^3}, \quad \frac{a}{bc^4}, \quad \frac{a}{bc^5}.$$

$$s_n = \left(\frac{a}{b}\right)\left(\frac{1}{c}\right)^{n-1}$$

10. $a = -3$; $r = \frac{3}{2} \div \frac{-3}{1} = \frac{-1}{2}$; $s_n = (-3)\left(\frac{-1}{2}\right)^{n-1}$

The eighth term is $s_8 = (-3)\left(\frac{-1}{2}\right)^7 = \frac{3}{128}$.

12. $a = -81a$; $r = \frac{-27a^2}{-81a} = \frac{1}{3}a$; $s_n = (-81a)\left(\frac{1}{3}a\right)^{n-1}$

The ninth term is $s_9 = (-81a)\left(\frac{1}{3}a\right)^8 = (-3^4a)\left(\frac{1}{3}a\right)^8$

$$= \frac{-1}{81}a^9.$$

14. $s_5 = 1$; $r = -\frac{1}{2}$, $n = 5$.

$s_n = ar^{n-1}$; $1 = a\left(-\frac{1}{2}\right)^4$; $a = 16$.

16. $\underline{-4}$, r $\underline{?}$, r $\underline{?}$, r $\underline{-32}$. There are 3 multiplications by r between -4 and -32. Hence, $r^3 = \frac{-32}{-4} = 8$; $r = 2$. Therefore, from two successive multiplications by 2:

$$-4 \cdot 2 = -8 \quad \text{and} \quad -8 \cdot 2 = -16.$$

The required geometric means are -8 and -16.

18. $\underline{-12}$, r $\underline{\quad}$, r $\underline{\frac{-1}{12}}$. There are 2 multiplications by r between -12 and $-\frac{1}{12}$. Hence,

$$r^2 = \frac{\frac{-1}{12}}{-12} = \frac{1}{144}; \quad r = \pm\frac{1}{12}.$$

Therefore, the required geometric mean is $-12 \cdot \left(\pm\frac{1}{12}\right) = \pm 1$.

20. $\underline{-25}$, r $\underline{?}$, r $\underline{?}$, r $\underline{?}$, r $\underline{\frac{-1}{25}}$. There are 4 multiplications by r between -25 and $-\frac{1}{25}$. Hence,

$$r^4 = \frac{\frac{-1}{25}}{-25} = \frac{1}{625}; \quad r = \frac{\pm 1}{5}.$$

With $r = \frac{1}{5}$, the geometric means are -5, -1, and $\frac{-1}{5}$. With $r = \frac{-1}{5}$, the geometric means are 5, -1, and $\frac{1}{5}$.

22. $\sum_{j=1}^{4} (-2)^j = (-2) + (-2)^2 + (-2)^3 + (-2)^4$; $a = -2$, $r = -2$, $n = 4$.

$$s_n = \frac{(-2) - (-2)(-2)^4}{1 - (-2)} = \frac{(-2) + 2(16)}{3} = 10$$

24. $\sum_{i=3}^{12} (2)^{i-5} = 2^{-2} + 2^{-1} + \cdots + 2^7$, $a = 2^{-2} = \frac{1}{4}$; $r = 2$, $n = 12 - 2 = 10$.

$$s_{10} = \frac{\frac{1}{4} - \frac{1}{4}(2)^{10}}{1 - 2} = \frac{\frac{1}{4} - 256}{-1} = \frac{4\left(\frac{1}{4} - 256\right)}{4(-1)} = \frac{1023}{4}$$

26. $\sum_{k=1}^{5} \left(\frac{1}{4}\right)^k = \left(\frac{1}{4}\right) + \left(\frac{1}{4}\right)^2 + \cdots + \left(\frac{1}{4}\right)^5$; $a = \frac{1}{4}$, $r = \frac{1}{4}$, $n = 5$.

$$s_5 = \frac{\frac{1}{4} - \frac{1}{4}\left(\frac{1}{4}\right)^5}{1 - \frac{1}{4}} = \frac{\frac{4^6}{1}\left(\frac{1}{4} - \frac{1}{4^6}\right)}{\frac{4^6}{1}\left(\frac{3}{4}\right)} = \frac{4^5 - 1}{4^5 \cdot 3}$$

28. After 2 hours: 40
After 4 hours: 160

The number of bacteria at the end of each hour is a geometric progression: 20, 40, 80, 160, . . . , with $a = 20$ and $r = 2$. Hence,

$$s_n = 20(2)^{n-1} = 5 \cdot 2^2(2^{n-1}) = 5 \cdot 2^{n+1}$$

30. After 2400 years: 50 grams; after 4800 years: 25 grams; after 7200 years: 12.5 grams; after 9600 years: 6.25 grams.

EXERCISE 11.4

In the following problems, we apply the formula for the sum of an infinite geometric progression: $S_\infty = \dfrac{a}{1 - r}$.

2. $a = 2, \quad r = \dfrac{1}{2};$
   $|r| < 1;$ hence, S_∞ exists.
   $$S_\infty = \frac{2}{1 - \frac{1}{2}} = 4$$

4. $a = 1, \quad r = \dfrac{2}{3};$
   $|r| < 1;$ hence, S_∞ exists.
   $$S_\infty = \frac{1}{1 - \frac{2}{3}} = \frac{1}{\frac{1}{3}} = 3$$

6. $r = \dfrac{-1}{8} \div \dfrac{1}{16} = -2;$
   $|r| = 2 > 1;$ hence, S_∞ does not exist.

8. $r = \dfrac{-3}{2} \div \dfrac{2}{1} = \dfrac{-3}{4}$
   $|r| = \dfrac{3}{4} < 1;$ hence, S_∞ exists. $a = 2;$
   $$S_\infty = \frac{2}{1 - \left(\frac{-3}{4}\right)} = \frac{2}{\frac{7}{4}} = \frac{8}{7}$$

10. $\dfrac{-1}{4} + \left(\dfrac{-1}{4}\right)^2 + \left(\dfrac{-1}{4}\right)^3 + \cdots;$
    $$a = \frac{-1}{4}, \quad r = \frac{-1}{4};$$
    $$S_\infty = \frac{\frac{-1}{4}}{1 - \left(\frac{-1}{4}\right)} = \frac{-1}{5}$$

12. $0.666\bar{6} = 0.6 + 0.06 + 0.006 + \cdots$
    $= \dfrac{6}{10} + \dfrac{6}{10^2} + \dfrac{6}{10^3} + \cdots;$
    $$a = \frac{6}{10}, \quad r = \frac{1}{10};$$
    $$S_\infty = \frac{\frac{6}{10}}{1 - \frac{1}{10}} = \frac{2}{3}$$

14. $0.454\overline{545} = 0.45 + 0.0045 + 0.000045 + \cdots$

$$= \frac{45}{100} + \frac{45}{10,000} + \frac{45}{1,000,000} + \cdots;$$

$$a = \frac{45}{100}, \quad r = \frac{1}{100};$$

$$S_\infty = \frac{\frac{45}{100}}{1 - \frac{1}{100}} = \frac{5}{11}$$

16. $3.027\overline{027} = 3 + 0.027 + 0.000027 + \cdots$

$$= 3 + \frac{27}{1,000} + \frac{27}{1,000,000} + \cdots;$$

$$a = \frac{27}{1,000}, \quad r = \frac{1}{1,000};$$

$$S_\infty = \frac{\frac{27}{1,000}}{1 - \frac{1}{1,000}} = \frac{1}{37}; \quad \text{hence,} \quad 3.027\overline{027} = 3\frac{1}{37}.$$

18. $0.833\bar{3} = 0.8 + 0.03 + 0.003 + 0.0003 + \cdots$

$$= \frac{8}{10} + \frac{3}{100} + \frac{3}{1,000} + \frac{3}{10,000} + \cdots; \quad a = \frac{3}{100},$$

$$r = \frac{1}{10};$$

$$S_\infty = \frac{\frac{3}{100}}{1 - \frac{1}{10}} = \frac{1}{30}; \quad \text{hence,}$$

$$0.833\bar{3} = \frac{8}{10} + \frac{1}{30} = \frac{5}{6}.$$

20. In the progression: $12 + 12\left(\frac{9}{10}\right) + 12\left(\frac{9}{10}\right)^2 + \cdots;$

$a = 12, \quad r = \frac{9}{10};$

$S_\infty = \frac{12}{1 - \frac{9}{10}} = 120.$ The pendulum will swing approximately 120 inches.

EXERCISE 11.5

2. For $n = 4$,

$$(3n)! = 12!$$
$$= 12 \cdot 11 \cdot 10 \cdot 9 \cdot 8 \cdot 7 \cdot 6 \cdot 5 \cdot 4 \cdot 3 \cdot 2 \cdot 1.$$

4. For $n = 4$
$$3n! = 3 \cdot 4! = 3(4 \cdot 3 \cdot 2 \cdot 1).$$

6. For $n = 2$
$$2n(2n - 1)! = 4 \cdot 3! = 4(3 \cdot 2 \cdot 1).$$

8. $7 \cdot 6 \cdot 5 \cdot 4 \cdot 3 \cdot 2 \cdot 1 = 5040$

10. $\dfrac{12 \cdot 11!}{11!} = 12$

12. $$\frac{(12!)(8!)}{16 \cdot 15 \cdot 14 \cdot 13(12!)} = \frac{8 \cdot 7 \cdot 6 \cdot 5 \cdot 4 \cdot 3 \cdot 2 \cdot 1}{16 \cdot 15 \cdot 14 \cdot 13} = \frac{12}{13}$$

14. $$\frac{10 \cdot 9 \cdot 8 \cdot 7 \cdot 6!}{4!6!} = \frac{10 \cdot 9 \cdot 8 \cdot 7}{4 \cdot 3 \cdot 2 \cdot 1} = 210$$

16. $5!$

18. $\dfrac{(7) \cdot 6!}{6!} = \dfrac{7!}{6!}$

20. $$\frac{(28 \cdot 27 \cdot 26 \cdot 25 \cdot 24) \cdot 23!}{23!} = \frac{28!}{23!}$$

22. $(n + 4)(n + 3)(n + 2) \cdot \cdots \cdot 3 \cdot 2 \cdot 1$

24. $3 \cdot n \cdot (n - 1) \cdot (n - 2) \cdot \cdots \cdot 3 \cdot 2 \cdot 1$

26. $(3n - 2)(3n - 3)(3n - 4) \cdot \cdots \cdot 3 \cdot 2 \cdot 1$

28. $$(2x)^4 + 4(2x)^3y + \frac{4 \cdot 3}{1 \cdot 2}(2x)^2y^2 + \frac{4 \cdot 3 \cdot 2}{1 \cdot 2 \cdot 3}(2x)y^3 + y^4$$
$$= 16x^4 + 32x^3y + 24x^2y^2 + 8xy^3 + y^4$$

30. $$(2x)^5 + 5(2x)^4(-1) + \frac{5 \cdot 4}{1 \cdot 2}(2x)^3(-1)^2 + \frac{5 \cdot 4 \cdot 3}{1 \cdot 2 \cdot 3}(2x)^2(-1)^3$$
$$+ \frac{5 \cdot 4 \cdot 3 \cdot 2}{1 \cdot 2 \cdot 3 \cdot 4}(2x)(-1)^4 + (-1)^5$$
$$= 32x^5 - 80x^4 + 80x^3 - 40x^2 + 10x - 1$$

32. $$\left(\frac{x}{3}\right)^5 + 5\left(\frac{x}{3}\right)^4(3) + \frac{5 \cdot 4}{1 \cdot 2}\left(\frac{x}{3}\right)^3(3)^2 + \frac{5 \cdot 4 \cdot 3}{1 \cdot 2 \cdot 3}\left(\frac{x}{3}\right)^2(3)^3 + \frac{5 \cdot 4 \cdot 3 \cdot 2}{1 \cdot 2 \cdot 3 \cdot 4}\left(\frac{x}{3}\right)(3)^4 + 3^5$$

$$= \frac{x^5}{243} + \frac{5x^4}{27} + \frac{10x^3}{3} + 30x^2 + 135x + 243$$

34. $$\left(\frac{2}{3}\right)^4 + 4\left(\frac{2}{3}\right)^3(-a^2) + \frac{4 \cdot 3}{1 \cdot 2}\left(\frac{2}{3}\right)^2(-a^2)^2 + \frac{4 \cdot 3 \cdot 2}{1 \cdot 2 \cdot 3}\left(\frac{2}{3}\right)(-a^2)^3 + (-a^2)^4$$

$$= \frac{16}{81} - \frac{32a^2}{27} + \frac{8a^4}{3} - \frac{8a^6}{3} + a^8$$

36. $$x^{15} + 15x^{14}(-y) + \frac{15 \cdot 14}{2!}x^{13}(-y)^2 + \frac{15 \cdot 14 \cdot 13}{3!}x^{12}(-y)^3$$

38. $$(2a)^{12} + 12(2a)^{11}(-b) + \frac{12 \cdot 11}{2!}(2a)^{10}(-b)^2 + \frac{12 \cdot 11 \cdot 10}{3!}(2a)^9(-b)^3$$

40. $$\left(\frac{x}{2}\right)^8 + 8\left(\frac{x}{2}\right)^7(2) + \frac{8 \cdot 7}{2!}\left(\frac{x}{2}\right)^6(2)^2 + \frac{8 \cdot 7 \cdot 6}{3!}\left(\frac{x}{2}\right)^5(2)^3$$

42. In Formula (2) of this section, substitute 12 for n and 5 for r:

$$\frac{12 \cdot 11 \cdot 10 \cdot 9}{(5 - 1)!}x^{12-5+1}(2)^{5-1} = 7920x^8.$$

44. In Formula (2) of this section, substitute 9 for n and 7 for r:

$$\frac{9 \cdot 8 \cdot 7 \cdot 6 \cdot 5 \cdot 4}{(7 - 1)!}(a^3)^{9-7+1}(-b)^{7-1} = 84a^9b^6.$$

46. In Formula (2) of this section, substitute 8 for n and 4 for r:

$$\frac{8 \cdot 7 \cdot 6}{(4 - 1)!}x^{8-4+1}\left(-\frac{1}{2}\right)^{4-1} = -7x^5$$

48. In Formula (2) of this section, substitute 10 for n and 8 for r:

$$\frac{10 \cdot 9 \cdot 8 \cdot 7 \cdot 6 \cdot 5 \cdot 4}{(8 - 1)!}\left(\frac{x}{2}\right)^{10-8+1}(4)^{8-1} = 245760x^3.$$

50. a. $(1 + 0.02)^{1/2}$

$$= (1)^{1/2} + \frac{1}{2}(1)^{-1/2}(0.02) + \frac{\left(\frac{1}{2}\right)\left(\frac{-1}{2}\right)}{2!}(1)^{-3/2}(0.02)^2 + \cdots$$

$$= 1 + 0.01 - 0.00005 + \cdots \approx 1.01$$

b. $(1 - 0.01)^{1/2}$

$$= (1)^{1/2} + \frac{1}{2}(1)^{-1/2}(-0.01) + \frac{\left(\frac{1}{2}\right)\left(\frac{-1}{2}\right)}{2!}(1)^{-3/2}(0.01)^2 + \cdots$$

$$= 1 - 0.005 - 0.0000125 \approx 0.99$$

EXERCISE 11.6

2. $p(7,7) = 7! = 5040$

4. $p(10,10) = 10! = 3,628,800$

6. The first digit cannot be zero, and the last digit must be one of three odd digits. $\underline{6} \cdot \underline{7} \cdot \underline{7} \cdot \underline{3} = 882$

8. There are seven numbers, three of which are odd, so there are three ways of picking the last digit (1, 3, or 5). Once it has been picked and since zero cannot be used as the first digit, there are five choices for the first digit. Now two numbers have been picked, leaving five choices for the second digit; and, with the second digit chosen, four choices remain for the third digit.

$$\underline{5} \cdot \underline{5} \cdot \underline{4} \cdot \underline{3} = 300$$

10. $\underline{8} \cdot \underline{9} \cdot \underline{4} = 288$ (see Problem 6)

12. $\underline{7} \cdot \underline{7} \cdot \underline{4} = 196$ (see Problem 8)

14. $\underline{26} \cdot \underline{26} \cdot \underline{26} \cdot \underline{26} \cdot \underline{26} = 11,881,376$

16. $\underline{10} \cdot \underline{26} \cdot \underline{10} \cdot \underline{10} = 26,000$

or if zero is not allowable in the first position,

$\underline{9} \cdot \underline{26} \cdot \underline{10} \cdot \underline{10} = 23,400.$

18. $$\frac{6!}{2!} = \frac{6 \cdot 5 \cdot 4 \cdot 3 \cdot 2 \cdot 1}{2 \cdot 1} = 360$$

20. $$\frac{7!}{3!} = \frac{7 \cdot 6 \cdot 5 \cdot 4 \cdot 3 \cdot 2 \cdot 1}{3 \cdot 2 \cdot 1} = 840$$

22. $\frac{8!}{3!} = 8 \cdot 7 \cdot 6 \cdot 5 \cdot 4 = 6720$

24. $\frac{11!}{2!} = 11 \cdot 10 \cdot 9 \cdot 8 \cdot 7 \cdot 6 \cdot 5 \cdot 4 \cdot 3 = 19{,}958{,}400$

26. $\frac{9!}{4!2!2!} = \frac{9 \cdot 8 \cdot 7 \cdot 6 \cdot 5}{2 \cdot 1 \cdot 2 \cdot 1} = 3780$

28. $\frac{11!}{3!2!2!2!} = \frac{11 \cdot 10 \cdot 9 \cdot 8 \cdot 7 \cdot 6 \cdot 5 \cdot 4}{2 \cdot 1 \cdot 2 \cdot 1 \cdot 2 \cdot 1} = 831{,}600$

EXERCISE 11.7

2. $\binom{9}{5} = \frac{9!}{5!4!} = \frac{9 \cdot 8 \cdot 7 \cdot 6}{4 \cdot 3 \cdot 2 \cdot 1} = 126$

4. 1 coin: $\binom{5}{1} = 5;$

   2 coins: $\binom{5}{2} = \frac{5 \cdot 4}{2 \cdot 1} = 10;$

   3 coins: $\binom{5}{3} = \frac{5 \cdot 4 \cdot 3}{3 \cdot 2 \cdot 1} = 10;$

   4 coins: $\binom{5}{4} = \frac{5 \cdot 4 \cdot 3 \cdot 2}{4 \cdot 3 \cdot 2 \cdot 1} = 5;$

   5 coins: $\binom{5}{5} = 1$ amount.

   Therefore, the total number of different amounts is

$$\binom{5}{1} + \binom{5}{2} + \binom{5}{3} + \binom{5}{4} + \binom{5}{5} = 31.$$

6. $\binom{52}{13} = \frac{52!}{13!39!} = 635{,}013{,}559{,}600$

8. An octagon has eight vertices, no three of which are collinear. The number of ways of joining pairs of these eight points is $\binom{8}{2} = 28$. Since eight of these connections will be the sides of the octagon, that leaves $28 - 8 = 20$ diagonals.

10. Two face cards can be drawn from the 12 face cards in a standard deck in $\binom{12}{2}$ ways.

$$\binom{12}{2} = \frac{12!}{2!10!} = \frac{12 \cdot 11 \cdot 10!}{2 \cdot 1 \cdot 10!} = 66$$

12. There are four aces; hence, an ace can be drawn in $\binom{4}{4} = 1$ way. There are forty-eight cards that are not aces; hence, a card that is not an ace can be drawn in $\binom{48}{1} = 48$ ways. Therefore, four aces can be paired with a card that is not an ace to form the specified hand in $1 \cdot 48 = 48$ ways.

14. A committee of eight men can be selected in $\binom{10}{8} = 45$ ways. A committee of eight women can be selected in $\binom{11}{8} = 165$ ways. For each way a committee of eight men can be chosen, there are 165 ways a committee of eight women can be chosen. Hence, a committee of eight men and eight women can be chosen in $45 \cdot 165 = 7425$ ways.

16. $\binom{100}{98} = \binom{100}{2} = \dfrac{100 \cdot 99}{2!} = 4950$

18. Using the results of Problem 15, if $r = 7$, then

$$n - r = 5; \quad n - 7 = 5; \quad n = 12.$$

20. The first five terms of $(a + b)^n$ are:

$$a^n + \frac{n}{1!}a^{n-1}b + \frac{n(n-1)}{2!}a^{n-2}b^2 + \frac{n(n-1)(n-2)}{3!}a^{n-3}b^3 + \frac{n(n-1)(n-2)(n-3)}{4!}a^{n-4}b^4 .$$

Then $\binom{n}{0} = \dfrac{n!}{0!n!} = 1;\quad \binom{n}{1} = \dfrac{n!}{1!(n-1)!} = \dfrac{n(n-1)!}{1!(n-1)!} = \dfrac{n}{1!};$

$$\binom{n}{2} = \frac{n!}{2!(n-2)!} = \frac{n(n-1)(n-2)!}{2!(n-2)!} = \frac{n(n-1)}{2!};$$

$$\binom{n}{3} = \frac{n!}{3!(n-3)!} = \frac{n(n-1)(n-2)(n-3)!}{3!(n-3)!}$$

$$= \frac{n(n-1)(n-2)}{3!}$$

$$\binom{n}{4} = \frac{n!}{4!(n-4)!} = \frac{n(n-1)(n-2)(n-3)(n-4)!}{4!(n-4)!}$$

$$= \frac{n(n-1)(n-2)(n-3)}{4!}$$

Therefore, the first five terms of the expansion of $(a + b)^n$ can be written as

$$\binom{n}{0}a^n + \binom{n}{1}a^{n-1}b + \binom{n}{2}a^{n-2}b^2 + \binom{n}{3}a^{n-3}b^3 + \binom{n}{4}a^{n-4}b^4.$$

APPENDIX A

SYNTHETIC DIVISION; POLYNOMIAL FUNCTIONS

EXERCISE A.1

2. $$\begin{array}{r|rrr} -3 & 1 & 1 & -6 \\ & & -3 & 6 \\ \hline & 1 & -2 & 0 \end{array}$$ $a - 2 \quad (a \neq -3)$

4. $$\begin{array}{r|rrr} -3 & 1 & 6 & 9 \\ & & -3 & -9 \\ \hline & 1 & 3 & 0 \end{array}$$ $x + 3 \quad (x \neq -3)$

6. $$\begin{array}{r|rrrrr} 3 & 1 & 0 & 2 & -3 & 5 \\ & & 3 & 9 & 33 & 90 \\ \hline & 1 & 3 & 11 & 30 & 95 \end{array}$$ $x^3 + 3x^2 + 11x + 30 + \frac{95}{x - 3} \quad (x \neq 3)$

8. $$\begin{array}{r|rrrr} -2 & 3 & 1 & 0 & -7 \\ & & -6 & 10 & -20 \\ \hline & 3 & -5 & 10 & -27 \end{array}$$ $3x^2 - 5x + 10 + \frac{-27}{x + 2} \quad (x \neq -2)$

10. $$\begin{array}{r|rrrrr} 4 & 3 & 0 & -1 & 0 & 1 \\ & & 12 & 48 & 188 & 752 \\ \hline & 3 & 12 & 47 & 188 & 753 \end{array}$$ $3x^3 + 12x^2 + 47x + 188 + \frac{753}{x - 4}$ $(x \neq 4)$

12. $$\begin{array}{r|rrrr} -3 & 1 & -7 & -1 & 3 \\ & & -3 & 30 & -87 \\ \hline & 1 & -10 & 29 & -84 \end{array}$$ $x^2 - 10x + 29 + \frac{-84}{x + 3} \quad (x \neq -3)$

14.

2	1	0	0	3	0	-2	-1
		2	4	8	22	44	84
	1	2	4	11	22	42	83

$x^5 + 2x^4 + 4x^3 + 11x^2 + 22x + 42 + \frac{83}{x - 2} \quad (x \neq 2)$

16.

-1	1	0	0	0	0	1
		-1	1	-1	1	-1
	1	-1	1	-1	1	0

$x^4 - x^3 + x^2 - x + 1 \quad (x \neq -1)$

18.

-1	1	0	0	0	0	0	1
		-1	1	-1	1	-1	1
	1	-1	1	-1	1	-1	2

$x^5 - x^4 + x^3 - x^2 + x - 1 + \frac{2}{x + 1} \quad (x \neq -1)$

EXERCISE A.2

2.

1	4	-2	3	0	-5
		4	2	5	5
	4	2	5	5	0

$P(1) = 0$

-1	4	-2	3	0	-5
		-4	6	-9	9
	4	-6	9	-9	4

$P(-1) = 4$

4.

-2	1	-10	5	-3	6
		-2	24	-58	122
	1	-12	29	-61	128

$P(-2) = 128$

3	1	-10	5	-3	6
		3	-21	-48	-153
	1	-7	-16	-51	-147

$P(3) = -147$

Note for Problems 6-12:

When $P(x)$ is divided by $x - a$, the remainder term is $P(a)$. Thus, synthetic division is a convenient means of obtaining ordered pairs $(a, P(a))$, which can be used to approximate the graph of $y = P(x)$. Usually, arbitrary integral values of a are used; however, to better approximate high or low points, fractional values are sometimes selected.

6.

a						$P(a)$	$(a,P(a))$
-5	-5	1	5	4	0		
			-5	0	-20		
		1	0	4	-20	-20	(-5,-20)
-4	-4	1	5	4	0		
			-4	-4	0		
		1	1	0	0	0	(-4,0)

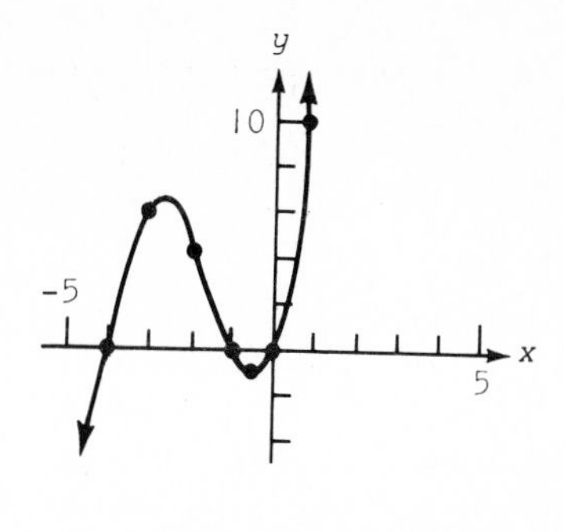

Continuing in the above manner, the following additional ordered pairs are obtained: (-3,6), (-2,4), (-1,0), (-0.5,-0.875), (0,0), (1,10).

8.

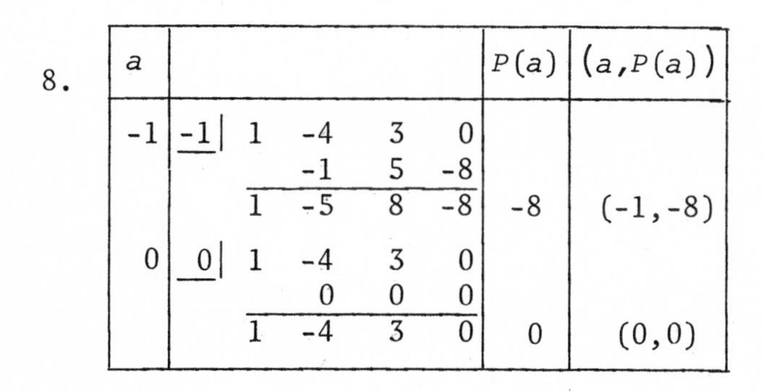

a						$P(a)$	$(a,P(a))$
-1	-1	1	-4	3	0		
			-1	5	-8		
		1	-5	8	-8	-8	(-1,-8)
0	0	1	-4	3	0		
			0	0	0		
		1	-4	3	0	0	(0,0)

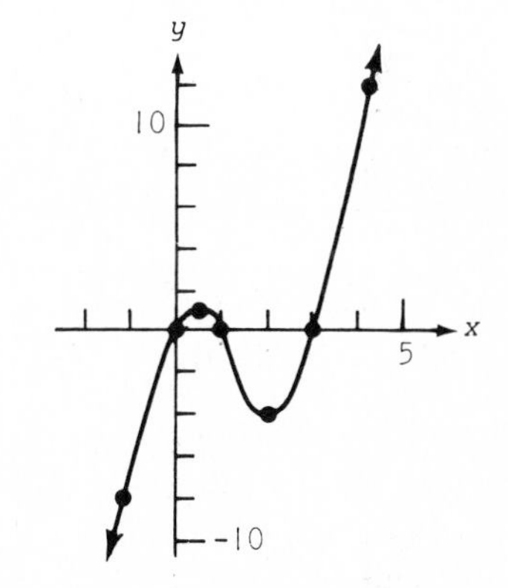

Continuing in the above manner, the following additional ordered pairs are obtained: (0.5,0.625), (1,0), (2,-2), (3,0), (4,12).

10.

a						$P(a)$	$(a,P(a))$
-3	-3	1	-3	-6	8		
			-3	18	-36		
		1	-6	12	-28	-28	(-3,-28)
-2	-2	1	-3	-6	8		
			-2	10	-8		
		1	-5	4	0	0	(-2,0)

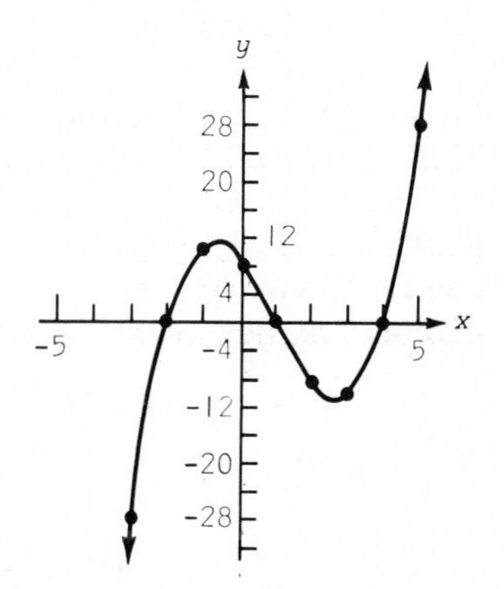

Continuing in the above manner, the following additional ordered pairs are obtained: (-1,10), (0,8), (1,0), (2,-8), (3,-10), (4,0), (5,28).

12.

a		$P(a)$	$(a,P(a))$
-3	$\underline{-3}\vert$ 1 -1 -4 4 0 -3 12 -24 60 1 -4 8 -20 60	60	(-3,60)
-2	$\underline{-2}\vert$ 1 -1 -4 4 0 -2 6 -4 0 1 -3 2 0 0	0	(-2,0)

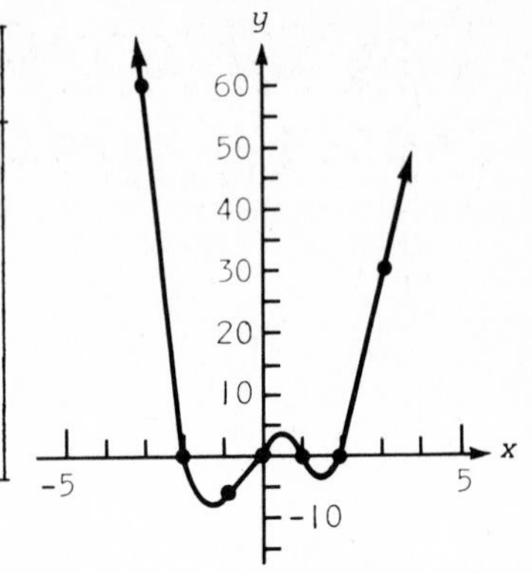

Continuing in the above manner, the following additional ordered pairs are obtained: (-1,-6), (0,0), (1,0), (2,0), (3,30).

14.
```
1| 2  -5   4  -1
       2  -3   1
  ---------------
   2  -3   1   0
```

Since $P(1) = 0$, by the factor theorem $x - 1$ is a factor.

16.
```
-1| 2  -5   3   3
       -2   7 -10
   ---------------
    2  -7  10  -7
```

Since $P(-1) \neq 0$, by the factor theorem $x + 1$ is not a factor.

18.
```
3| 1  -6  -1  30
       3  -9 -30
  ---------------
   1  -3 -10   0
```

$x^3 - 6x^2 - x + 30 = (x - 3)(x^2 - 3x - 10)$

Hence, $x^3 - 6x^2 - x + 30 = 0$ is equivalent to

$$(x - 3)(x^2 - 3x - 10) = 0$$
$$(x - 3)(x - 5)(x + 2) = 0.$$

By inspection, the solution set is $\{3,5,-2\}$.

20.
```
-5| 1   5  -1  -5   0
       -5   0   5   0
   -------------------
    1   0  -1   0   0
```

$x^4 + 5x^3 - x^2 - 5x = (x - (-5))(x^3 - x)$

Hence, $x^4 + 5x^3 - x^2 - 5x = 0$ is equivalent to

$$(x + 5)(x^3 - x) = 0$$
$$(x + 5)(x)(x^2 - 1) = 0$$
$$(x + 5)(x)(x - 1)(x + 1) = 0.$$

By inspection, the solution set is $\{-5,0,1,-1\}$.

APPENDIX B

MATRICES AND DETERMINANTS

EXERCISE B.1

2. The augmented matrix is

$$\begin{bmatrix} 1 & -5 & 11 \\ 2 & 3 & -4 \end{bmatrix}.$$

An equivalent matrix is

$$\text{row } 2 + [-2 \times \text{row } 1] \rightarrow \begin{bmatrix} 1 & -5 & 11 \\ 0 & 13 & -26 \end{bmatrix} \quad \begin{aligned} x - 5y &= 11 \\ 13y &= -26. \end{aligned}$$

Since, from the last equation, $y = -2$; -2 can be substituted for y in the first equation to obtain

$$x - 5(-2) = 11, \quad x = 1.$$

The solution set is $\{(1,-2)\}$.

4. The augmented matrix is

$$\begin{bmatrix} 1 & 6 & -14 \\ 5 & -3 & -4 \end{bmatrix}.$$

An equivalent matrix is

$$\text{row } 2 + [-5 \times \text{row } 1] \rightarrow \begin{bmatrix} 1 & 6 & -14 \\ 0 & -33 & 66 \end{bmatrix} \quad \begin{aligned} x + 6y &= -14 \\ -33y &= 66. \end{aligned}$$

From the last equation, $y = -2$; -2 can be substituted for y in the first equation to obtain

$$x + 6(-2) = -14, \quad x = -2.$$

The solution set is $\{(-2,-2)\}$.

6. The augmented matrix is

$$\begin{bmatrix} 3 & -2 & 16 \\ 4 & 2 & 12 \end{bmatrix}.$$

An equivalent matrix is

$$\text{row 1 + row 2} \rightarrow \begin{bmatrix} 3 & -2 & 16 \\ 7 & 0 & 28 \end{bmatrix} \quad \begin{aligned} 3x - 2y &= 16 \\ 7x \quad &= 28. \end{aligned}$$

From the last equation, $x = 4$; 4 can be substituted for x in the first equation to obtain

$$3(4) - 2y = 16, \quad y = -2.$$

The solution set is $\{(4,-2)\}$.

8. The augmented matrix is

$$\begin{bmatrix} 1 & -2 & 3 & -11 \\ 2 & 3 & -1 & 6 \\ 3 & -1 & -1 & 3 \end{bmatrix}.$$

An equivalent matrix is

$$\begin{matrix} \\ \text{row 2 + [-2} \times \text{row 1]} \rightarrow \\ \text{row 3 + [-3} \times \text{row 1]} \rightarrow \end{matrix} \begin{bmatrix} 1 & -2 & 3 & -11 \\ 0 & 7 & -7 & 28 \\ 0 & 5 & -10 & 36 \end{bmatrix}.$$

Equivalent to the above matrix is

$$\begin{matrix} \\ \\ \text{row 3} + \left[\frac{-5}{7} \times \text{row 2}\right] \rightarrow \end{matrix} \begin{bmatrix} 1 & -2 & 3 & -11 \\ 0 & 7 & -7 & 28 \\ 0 & 0 & -5 & 16 \end{bmatrix} \quad \begin{aligned} x - 2y + 3z &= -11 \\ 7y - 7z &= 28 \\ -5z &= 16. \end{aligned}$$

From the last equation, $z = \frac{-16}{5}$. In the second equation, substitute $\frac{-16}{5}$ for z and obtain

$$7y - 7\left(\frac{-16}{5}\right) = 28, \quad 7y + \frac{112}{5} = \frac{140}{5}, \quad y = \frac{4}{5}.$$

In the first equation, substitute $\frac{-16}{5}$ for z and $\frac{4}{5}$ for y to obtain

$$x - 2\left(\frac{4}{5}\right) + 3\left(\frac{-16}{5}\right) = -11, \quad x - \frac{8}{5} - \frac{48}{5} = \frac{-55}{5}, \quad x = \frac{1}{5}.$$

The solution set is $\left\{\left(\frac{1}{5}, \frac{4}{5}, \frac{-16}{5}\right)\right\}$.

10. The augmented matrix is

$$\begin{bmatrix} 1 & -2 & -2 & 4 \\ 2 & 1 & -3 & 7 \\ 1 & -1 & -1 & 3 \end{bmatrix}.$$

An equivalent matrix is

$$\begin{matrix} \\ \text{row } 2 + [-2 \times \text{row } 1] \rightarrow \\ \text{row } 3 + [-1 \times \text{row } 1] \rightarrow \end{matrix} \begin{bmatrix} 1 & -2 & -2 & 4 \\ 0 & 5 & 1 & -1 \\ 0 & 1 & 1 & -1 \end{bmatrix}.$$

Equivalent to the above matrix is

$$\begin{matrix} \\ \text{row } 2 + [-5 \times \text{row } 3] \rightarrow \\ \\ \end{matrix} \begin{bmatrix} 1 & -2 & -2 & 4 \\ 0 & 0 & -4 & 4 \\ 0 & 1 & 1 & -1 \end{bmatrix} \begin{matrix} x - 2y - 2z = 4 \\ -4z = 4 \\ y + z = -1. \end{matrix}$$

Note that in order to avoid introducing fractions, row 2 was chosen as the row to contain two zeros.

From the second equation, $z = -1$. In the third equation, substitute -1 for z and obtain

$$y + (-1) = -1, \quad y = 0.$$

In the first equation, substitute -1 for z and 0 for y and obtain

$$x - 2(0) - 2(-1) = 4, \quad x = 2.$$

The solution set is $\{(2,0,-1)\}$.

12. The augmented matrix is

$$\begin{bmatrix} 1 & -2 & -5 & 2 \\ 2 & 3 & 1 & 11 \\ 3 & -1 & -1 & 11 \end{bmatrix}.$$

An equivalent matrix is

$$\begin{matrix} \\ \text{row } 2 + [-2 \times \text{row } 1] \rightarrow \\ \text{row } 3 + [-3 \times \text{row } 1] \rightarrow \end{matrix} \begin{bmatrix} 1 & -2 & -5 & 2 \\ 0 & 7 & 11 & 7 \\ 0 & 5 & 14 & 5 \end{bmatrix}.$$

Equivalent to the above matrix is

$$\text{row } 3 + \left[\frac{-5}{7} \times \text{row } 2\right] \rightarrow \begin{bmatrix} 1 & -2 & -5 & 2 \\ 0 & 7 & 11 & 7 \\ 0 & 0 & \frac{43}{7} & 0 \end{bmatrix} \quad \begin{aligned} x - 2y - 5z &= 2 \\ 7y + 11z &= 7 \\ \frac{43}{7}z &= 0. \end{aligned}$$

From the last equation, $z = 0$. In the second equation, substitute 0 for z and obtain

$$7y + 11(0) = 7, \quad y = 1.$$

In the first equation, substitute 0 for z and 1 for y and obtain

$$x - 2(1) - 5(0) = 2, \quad x = 4.$$

The solution set is $\{(4,1,0)\}$.

EXERCISE B.2

2. $(3)(1) - (4)(-2) = 11$

4. $(1)(2) - (-1)(-2) = 0$

6. $(20)(-2) - (-20)(3) = 20$

8. $(-1)(-6) - (-2)(-5) = -4$

10. $D = \begin{vmatrix} 3 & -4 \\ 1 & -2 \end{vmatrix} = (3)(-2) - (1)(-4) = -2$

$$D_x = \begin{vmatrix} -2 & -4 \\ 0 & -2 \end{vmatrix} = (-2)(-2) - (0)(-4) = 4$$

$$D_y = \begin{vmatrix} 3 & -2 \\ 1 & 0 \end{vmatrix} = (3)(0) - (1)(-2) = 2$$

$$x = \frac{D_x}{D} = \frac{4}{-2} = -2; \qquad y = \frac{D_y}{D} = \frac{2}{-2} = -1$$

The solution set is $\{(-2,-1)\}$.

12. $D = \begin{vmatrix} 2 & -4 \\ 1 & -2 \end{vmatrix} = (2)(-2) - (1)(-4) = 0$

Since $D = 0$, the given equations are dependent or inconsistent and there is no unique solution.

14. $D = \begin{vmatrix} \frac{2}{3} & 1 \\ 1 & -\frac{4}{3} \end{vmatrix} = \left(\frac{2}{3}\right)\left(-\frac{4}{3}\right) - (1)(1) = \frac{-17}{9}$

$D_x = \begin{vmatrix} 1 & 1 \\ 0 & -\frac{4}{3} \end{vmatrix} = (1)\left(-\frac{4}{3}\right) - (0)(1) = \frac{-4}{3}$

$D_y = \begin{vmatrix} \frac{2}{3} & 1 \\ 1 & 0 \end{vmatrix} = \left(\frac{2}{3}\right)(0) - (1)(1) = -1$

$x = \frac{D_x}{D} = \frac{\frac{-4}{3}}{\frac{-17}{9}} = \frac{12}{17}; \qquad y = \frac{D_y}{D} = \frac{-1}{\frac{-17}{9}} = \frac{9}{17}$

The solution set is $\left\{\left(\frac{12}{17}, \frac{9}{17}\right)\right\}$.

16. $D = \begin{vmatrix} \frac{1}{2} & 1 \\ \frac{-1}{4} & -1 \end{vmatrix} = \left(\frac{1}{2}\right)(-1) - \left(\frac{-1}{4}\right)(1) = \frac{-1}{4}$

$D_x = \begin{vmatrix} 3 & 1 \\ -3 & -1 \end{vmatrix} = (3)(-1) - (-3)(1) = 0$

$D_y = \begin{vmatrix} \frac{1}{2} & 3 \\ \frac{-1}{4} & -3 \end{vmatrix} = \left(\frac{1}{2}\right)(-3) - \left(\frac{-1}{4}\right)(3) = \frac{-3}{4}$

$x = \frac{D_x}{D} = \frac{0}{\frac{-1}{4}} = 0; \qquad y = \frac{D_y}{D} = \frac{\frac{-3}{4}}{\frac{-1}{4}} = 3$

The solution set is $\{(0,3)\}$.

18. $D = \begin{vmatrix} 2 & -3 \\ 1 & 0 \end{vmatrix} = (2)(0) - (1)(-3) = 3$

$D_x = \begin{vmatrix} 12 & -3 \\ 4 & 0 \end{vmatrix} = (12)(0) - (4)(-3) = 12$

$$D_y = \begin{vmatrix} 2 & 12 \\ 1 & 4 \end{vmatrix} = (2)(4) - (1)(12) = -4$$

$$x = \frac{D_x}{D} = \frac{12}{3} = 4; \qquad y = \frac{D_y}{D} = \frac{-4}{3}$$

The solution set is $\left\{\left(4, \frac{-4}{3}\right)\right\}$.

20. $$D = \begin{vmatrix} 1 & 1 \\ 1 & -1 \end{vmatrix} = (1)(-1) - (1)(1) = -1 - 1 = -2$$

$$D_x = \begin{vmatrix} a & 1 \\ b & -1 \end{vmatrix} = (a)(-1) - (b)(1) = -a - b$$

$$D_y = \begin{vmatrix} 1 & a \\ 1 & b \end{vmatrix} = (1)(b) - (1)(a) = b - a$$

$$x = \frac{D_x}{D} = \frac{-a - b}{-2} = \frac{a + b}{2}; \qquad y \quad \frac{D_y}{D} = \frac{b - a}{-2} = \frac{a - b}{2}$$

The solution set is $\left\{\left(\frac{a + b}{2}, \frac{a - b}{2}\right)\right\}$.

22. $$-\begin{vmatrix} a_2 & b_2 \\ a_1 & b_1 \end{vmatrix} = -(a_2b_1 - a_1b_2)$$

$$\begin{vmatrix} a_1 & b_1 \\ a_2 & b_2 \end{vmatrix} = a_1b_2 - a_2b_1 = -(a_2b_1 - a_1b_2) = -\begin{vmatrix} a_2 & b_2 \\ a_1 & b_1 \end{vmatrix}$$

24. $$\begin{vmatrix} ka_1 & b_1 \\ ka_2 & b_2 \end{vmatrix} = ka_1b_2 - ka_2b_1 = k(a_1b_2 - a_2b_1) = k\begin{vmatrix} a_1 & b_1 \\ a_2 & b_2 \end{vmatrix}$$

26. $$\begin{vmatrix} a_1 + ka_2 & b_1 + kb_2 \\ a_2 & b_2 \end{vmatrix} = (a_1 + ka_2)b_2 - a_2(b_1 + kb_2)$$

$$= a_1b_2 + ka_2b_2 - a_2b_1 - ka_2b_2$$

$$= a_1b_2 - a_2b_1 = \begin{vmatrix} a_1 & b_1 \\ a_2 & b_2 \end{vmatrix}$$

28. $D = \begin{vmatrix} a_1 & b_1 \\ a_2 & b_2 \end{vmatrix} = a_1b_2 - a_2b_1$

$D_x = \begin{vmatrix} c_1 & b_1 \\ c_2 & b_2 \end{vmatrix} = c_1b_2 - c_2b_1$; $D_y = \begin{vmatrix} a_1 & c_1 \\ a_2 & c_2 \end{vmatrix} = a_1c_2 - a_2c_1$

If $D = 0$, then $a_1b_2 - a_2b_1 = 0$ or $a_1b_2 = a_2b_1$. Thus,

$$\frac{a_1}{a_2} = \frac{b_1}{b_2}$$

If $D_x = 0$, then $c_1b_2 - c_2b_1 = 0$ or $c_2b_1 = c_1b_2$. Thus,

$$\frac{b_1}{b_2} = \frac{c_1}{c_2}$$

Hence, $\frac{a_1}{a_2} = \frac{c_1}{c_2}$ or $a_1c_2 = a_2c_1$. Thus, $a_1c_2 - a_2c_1 = 0 = D_y$.

EXERCISE B.3

2. Expand about the third row:

$$0\begin{vmatrix} 3 & 1 \\ 2 & 1 \end{vmatrix} - 2\begin{vmatrix} 1 & 1 \\ -1 & 1 \end{vmatrix} + 0\begin{vmatrix} 1 & 3 \\ -1 & 2 \end{vmatrix} = 0 - 2(2) + 0 = -4.$$

4. Expand about the third row:

$$4\begin{vmatrix} 4 & -1 \\ 3 & 2 \end{vmatrix} - 0\begin{vmatrix} 2 & -1 \\ -1 & 2 \end{vmatrix} + 2\begin{vmatrix} 2 & 4 \\ -1 & 3 \end{vmatrix} = 4(11) + 0 + 2(10) = 64.$$

6. Expand about the first row:

$$1\begin{vmatrix} 1 & 2 \\ 3 & 4 \end{vmatrix} - 0\begin{vmatrix} 0 & 2 \\ 0 & 4 \end{vmatrix} + 0\begin{vmatrix} 0 & 1 \\ 0 & 3 \end{vmatrix} = 1(-2) + 0 + 0 = -2$$

8. Expand about the second column:

$$-1\begin{vmatrix} 3 & 6 \\ 5 & 10 \end{vmatrix} + 2\begin{vmatrix} 2 & 4 \\ 5 & 10 \end{vmatrix} - (-3)\begin{vmatrix} 2 & 4 \\ 3 & 6 \end{vmatrix} = -1(0) + 2(0) + 3(0) = 0.$$

10. Expand about the second row:

$$-0\begin{vmatrix} 3 & 1 \\ 2 & 1 \end{vmatrix} + 1\begin{vmatrix} 2 & 1 \\ -4 & 1 \end{vmatrix} - 0\begin{vmatrix} 2 & 3 \\ -4 & 2 \end{vmatrix} = 0 + 1(6) + 0 = 6.$$

12. Expand about the first row:

$$a\begin{vmatrix} 2 & 3 \\ 5 & 6 \end{vmatrix} - a\begin{vmatrix} 1 & 3 \\ 4 & 6 \end{vmatrix} + a\begin{vmatrix} 1 & 2 \\ 4 & 5 \end{vmatrix} = a(-3) - a(-6) + a(-3) = -3a + 6a - 3a = 0.$$

14. Expand about the first row:

$$0\begin{vmatrix} x & 0 \\ 0 & 0 \end{vmatrix} - 0\begin{vmatrix} 0 & 0 \\ x & 0 \end{vmatrix} + x\begin{vmatrix} 0 & x \\ x & 0 \end{vmatrix} = 0 + 0 + x(-x^2) = -x^3$$

16. Expand about the first column:

$$0\begin{vmatrix} 0 & a \\ a & 0 \end{vmatrix} - a\begin{vmatrix} a & b \\ a & 0 \end{vmatrix} + b\begin{vmatrix} a & b \\ 0 & a \end{vmatrix} = 0 - a(-ab) + b(a^2) = 2a^2b.$$

18. Expand about the first row:

$$0\begin{vmatrix} a & b \\ b & 0 \end{vmatrix} - b\begin{vmatrix} b & b \\ 0 & 0 \end{vmatrix} + 0\begin{vmatrix} b & a \\ 0 & b \end{vmatrix} = 0 - b(0) + 0(b^2) = 0.$$

20. Expand about the first row:

$$x^2\begin{vmatrix} -1 & 3 \\ 2 & 0 \end{vmatrix} - 0\begin{vmatrix} 2 & 3 \\ 3 & 0 \end{vmatrix} + 1\begin{vmatrix} 2 & -1 \\ 3 & 2 \end{vmatrix} = 1; \quad x^2(-6) + 0 + 1(7) = 1;$$

$x^2 = 1$; $x = \pm 1$.

The solution set is $\{-1,1\}$.

22. Expand about the first column:

$$x\begin{vmatrix} x & 1 \\ x & 0 \end{vmatrix} - 0\begin{vmatrix} 1 & 1 \\ x & 0 \end{vmatrix} + 0\begin{vmatrix} 1 & 1 \\ x & 1 \end{vmatrix} = -4; \quad x(-x) - 0 + 0 = -4;$$

$x^2 = 4$; $x = \pm 2$.

The solution set is $\{-2,2\}$.

24. Expand about the first row:

$$0\begin{vmatrix} b & c \\ e & f \end{vmatrix} - 0\begin{vmatrix} a & c \\ d & f \end{vmatrix} + 0\begin{vmatrix} a & b \\ d & e \end{vmatrix} = 0 - 0 + 0 = 0.$$

If each entry in a row of a third-order determinant is 0, then the determinant is zero.

26. Expand $\begin{vmatrix} 2 & 0 & 1 \\ 4 & 1 & -2 \\ 6 & 1 & 1 \end{vmatrix}$ about the second column:

$$-0\begin{vmatrix} 4 & -2 \\ 6 & 1 \end{vmatrix} + 1\begin{vmatrix} 2 & 1 \\ 6 & 1 \end{vmatrix} - 1\begin{vmatrix} 2 & 1 \\ 4 & -2 \end{vmatrix} = 0 + 1(-4) - 1(-8) = 4.$$

Expand $\begin{vmatrix} 1 & 0 & 1 \\ 2 & 1 & -2 \\ 3 & 1 & 1 \end{vmatrix}$ about the second column:

$$-0\begin{vmatrix} 2 & -2 \\ 3 & 1 \end{vmatrix} + 1\begin{vmatrix} 1 & 1 \\ 3 & 1 \end{vmatrix} - 1\begin{vmatrix} 1 & 1 \\ 2 & -2 \end{vmatrix} = 0 + 1(-2) - 1(-4) = 2.$$

Thus, $\begin{vmatrix} 2 & 0 & 1 \\ 4 & 1 & -2 \\ 6 & 1 & 1 \end{vmatrix} = 2\begin{vmatrix} 1 & 0 & 1 \\ 2 & 1 & -2 \\ 3 & 1 & 1 \end{vmatrix}$.

If a common factor is factored from each element of a column in a determinant, the resulting determinant multiplied by the common factor equals the original determinant.

EXERCISE B.4

2. $D = \begin{vmatrix} 2 & -6 & 3 \\ 3 & -2 & 5 \\ 4 & 5 & -2 \end{vmatrix} = -129$ $\quad D_x = \begin{vmatrix} -12 & -6 & 3 \\ -4 & -2 & 5 \\ 10 & 5 & -2 \end{vmatrix} = 0$

$D_y = \begin{vmatrix} 2 & -12 & 3 \\ 3 & -4 & 5 \\ 4 & 10 & -2 \end{vmatrix} = -258$ $\quad D_z = \begin{vmatrix} 2 & -6 & -12 \\ 3 & -2 & -4 \\ 4 & 5 & 10 \end{vmatrix} = 0$

$$x = \frac{D_x}{D} = 0; \quad y = \frac{D_y}{D} = 2; \quad z = \frac{D_z}{D} = 0.$$

The solution set is $\{(0,2,0)\}$.

4. $D = \begin{vmatrix} 2 & 0 & 5 \\ 4 & 3 & 0 \\ 0 & 3 & -4 \end{vmatrix} = 36$ $\quad D_x = \begin{vmatrix} 9 & 0 & 5 \\ -1 & 3 & 0 \\ -13 & 3 & -4 \end{vmatrix} = 72$

$D_y = \begin{vmatrix} 2 & 9 & 5 \\ 4 & -1 & 0 \\ 0 & -13 & -4 \end{vmatrix} = -108$ $\quad D_z = \begin{vmatrix} 2 & 0 & 9 \\ 4 & 3 & -1 \\ 0 & 3 & -13 \end{vmatrix} = 36$

$x = \frac{D_x}{D} = 2; \quad y = \frac{D_y}{D} = -3; \quad z = \frac{D_z}{D} = 1.$

The solution set is $\{(2,-3,1)\}$.

6. $D = \begin{vmatrix} 4 & 8 & 1 \\ 2 & -3 & 2 \\ 1 & 7 & -3 \end{vmatrix} = 61$ $\quad D_x = \begin{vmatrix} -6 & 8 & 1 \\ 0 & -3 & 2 \\ -8 & 7 & -3 \end{vmatrix} = -122$

$D_y = \begin{vmatrix} 4 & -6 & 1 \\ 2 & 0 & 2 \\ 1 & -8 & -3 \end{vmatrix} = 0$ $\quad D_z = \begin{vmatrix} 4 & 8 & -6 \\ 2 & -3 & 0 \\ 1 & 7 & -8 \end{vmatrix} = 122$

$x = \frac{D_x}{D} = -2; \quad y = \frac{D_y}{D} = 0; \quad z = \frac{D_z}{D} = 2.$

The solution set is $\{(-2,0,2)\}$.

8. $D = \begin{vmatrix} 1 & 1 & -2 \\ 3 & -1 & 1 \\ 3 & 3 & -6 \end{vmatrix} = 0$

Since $D = 0$, there is no unique solution.

10. $D = \begin{vmatrix} 3 & -2 & 5 \\ 4 & -4 & 3 \\ 5 & -4 & 1 \end{vmatrix} = 22$ $\quad D_x = \begin{vmatrix} 6 & -2 & 5 \\ 0 & -4 & 3 \\ -5 & -4 & 1 \end{vmatrix} = -22$

$D_y = \begin{vmatrix} 3 & 6 & 5 \\ 4 & 0 & 3 \\ 5 & -5 & 1 \end{vmatrix} = 11$ $\quad D_z = \begin{vmatrix} 3 & -2 & 6 \\ 4 & -4 & 0 \\ 5 & -4 & -5 \end{vmatrix} = 44$

$x = \frac{D_x}{D} = -1; \quad y = \frac{D_y}{D} = \frac{1}{2}; \quad z = \frac{D_z}{D} = 2.$

The solution set is $\left\{\left(-1,\frac{1}{2},2\right)\right\}$.

12. Multiply the first equation by 3 and the second equation by 12.

$$6x - 2y + 3z = 6$$
$$6x - 4y - 3z = 0$$
$$4x + 5y - 3z = -1$$

$$D = \begin{vmatrix} 6 & -2 & 3 \\ 6 & -4 & -3 \\ 4 & 5 & -3 \end{vmatrix} = 288 \qquad D_x = \begin{vmatrix} 6 & -2 & 3 \\ 0 & -4 & -3 \\ -1 & 5 & -3 \end{vmatrix} = 144$$

$$D_y = \begin{vmatrix} 6 & 6 & 3 \\ 6 & 0 & -3 \\ 4 & -1 & -3 \end{vmatrix} = 0 \qquad D_z = \begin{vmatrix} 6 & -2 & 6 \\ 6 & -4 & 0 \\ 4 & 5 & -1 \end{vmatrix} = 288$$

$$x = \frac{D_x}{D} = \frac{1}{2}; \quad y = \frac{D_y}{D} = 0; \quad z = \frac{D_z}{D} = 1.$$

The solution set is $\left\{\left(\frac{1}{2}, 0, 1\right)\right\}$.

14. $$D = \begin{vmatrix} 2 & 1 & 0 \\ 0 & 1 & 1 \\ 3 & -2 & -5 \end{vmatrix} = -3 \qquad D_x = \begin{vmatrix} 18 & 1 & 0 \\ -1 & 1 & 1 \\ 38 & -2 & -5 \end{vmatrix} = -21$$

$$D_y = \begin{vmatrix} 2 & 18 & 0 \\ 0 & -1 & 1 \\ 3 & 38 & -5 \end{vmatrix} = -12 \qquad D_z = \begin{vmatrix} 2 & 1 & 18 \\ 0 & 1 & -1 \\ 3 & -2 & 38 \end{vmatrix} = 15$$

$$x = \frac{D_x}{D} = 7; \quad y = \frac{D_y}{D} = 4; \quad z = \frac{D_z}{D} = -5.$$

The solution set is $\{(7,4,-5)\}$.

LINEAR INTERPOLATION

EXERCISE C.1

2. $\log_{10} 8.184 = \log_{10}(8.184 \times 10^0)$. Hence, the characteristic is 0.

$$10\left\{\begin{array}{l} 4\left\{\begin{array}{l} 8.180 \\ 8.184 \end{array}\right. \\ \phantom{4\{}\,8.190 \end{array}\right. \quad \begin{array}{l} \underline{\log_{10} x} \\ 0.9128 \\ \quad ? \\ 0.9133 \end{array} \left.\begin{array}{l} \left.\begin{array}{l} \\ \end{array}\right\} y \\ \\ \end{array}\right\} 0.0005$$

(column headings: x, $\log_{10} x$)

$$\frac{y}{0.0005} = \frac{4}{10}$$

$$y = \frac{4}{10}(0.0005) = 0.0002$$

$$\log_{10} 8.184 = 0.9128 + 0.0002 = 0.9130$$

4. $\log_{10} 10.31 = \log_{10}(1.031 \times 10^1)$. Hence, the characteristic is 1.

$$10\left\{\begin{array}{l} 1\left\{\begin{array}{l} 1.030 \\ 1.031 \end{array}\right. \\ \phantom{1\{}\,1.040 \end{array}\right. \quad \begin{array}{l} \underline{\log_{10} x} \\ 0.0128 \\ \quad ? \\ 0.0170 \end{array} \left.\begin{array}{l} \left.\begin{array}{l} \\ \end{array}\right\} y \\ \\ \end{array}\right\} 0.0042$$

(column headings: x, $\log_{10} x$)

$$\frac{y}{0.0042} = \frac{1}{10}$$

$$y = \frac{1}{10}(0.0042) \approx 0.0004$$

$$\log_{10} 1.031 = 0.0128 + 0.0004 = 0.0132$$

Adding the characteristic,

$$\log_{10} 10.31 = 1 + 0.0132 = 1.0132.$$

6. $\log_{10} 203.4 = \log_{10}(2.034 \times 10^2)$. Hence, the characteristic is 2.

$$10\left\{\begin{array}{l} 4\left\{\begin{array}{l} 2.030 \\ 2.034 \end{array}\right. \\ \quad 2.040 \end{array}\right. \quad \left.\begin{array}{r} \left.\begin{array}{r} 0.3075 \\ ? \end{array}\right\} y \\ 0.3096 \quad \end{array}\right\} 0.0021$$

(columns: x, $\log_{10} x$)

$$\frac{y}{0.0021} = \frac{4}{10}$$

$$y = \frac{4}{10}(0.0021) \approx 0.0008$$

$$\log_{10} 2.034 = 0.3075 + 0.0008 = 0.3083$$

Adding the characteristic,

$$\log_{10} 203.4 = 2 + 0.3083 = 2.3083.$$

8. $\log_{10} 72.36 = \log_{10}(7.236 \times 10^1)$. Hence, the characteristic is 1.

$$10\left\{\begin{array}{l} 6\left\{\begin{array}{l} 7.230 \\ 7.236 \end{array}\right. \\ \quad 7.240 \end{array}\right. \quad \left.\begin{array}{r} \left.\begin{array}{r} 0.8591 \\ ? \end{array}\right\} y \\ 0.8597 \quad \end{array}\right\} 0.0006$$

(columns: x, $\log_{10} x$)

$$\frac{y}{0.0006} = \frac{6}{10}$$

$$y = \frac{6}{10}(0.0006) \approx 0.0004$$

$$\log_{10} 7.236 = 0.8591 + 0.0004 = 0.8595$$

Adding the characteristic,

$$\log_{10} 72.36 = 1 + 0.8595 = 1.8595.$$

10. $\log_{10} 0.09142 = \log_{10}(9.142 \times 10^{-2})$. Hence, the characteristic is -2 or 8 - 10.

$$10\left\{\begin{array}{l} 2\left\{\begin{array}{l} 9.140 \\ 9.142 \end{array}\right. \\ \quad 9.150 \end{array}\right. \quad \left.\begin{array}{r} \left.\begin{array}{r} 0.9609 \\ ? \end{array}\right\} y \\ 0.9614 \quad \end{array}\right\} 0.0005$$

(columns: x, $\log_{10} x$)

$$\frac{y}{0.0005} = \frac{2}{10}$$

$$y = \frac{2}{10}(0.0005) = 0.0001$$

$$\log_{10} 9.142 = 0.9609 + 0.0001 = 0.9610$$

Adding the characteristic,

$$\log_{10} 0.09142 = 8 - 10 + 0.9610 = 8.9610 - 10.$$

12. $\log_{10} 0.03741 = \log_{10}(3.741 \times 10^{-2})$. Hence, the characteristic is -2 or 8 - 10.

$$10\left\{\begin{array}{l} 1\left\{\begin{array}{l} 3.740 \\ 3.741 \end{array}\right. \\ \quad 3.750 \end{array}\right. \quad \left.\begin{array}{r} \left.\begin{array}{r} 0.5729 \\ ? \end{array}\right\} y \\ 0.5740 \quad \end{array}\right\} 0.0011$$

(columns: x, $\log_{10} x$)

$$\frac{y}{0.0011} = \frac{1}{10}$$

$$y = \frac{1}{10}(0.0011) \approx 0.0001$$

(continued)

$$\log_{10} 3.741 = 0.5729 + 0.0001 = 0.5730$$

Adding the characteristic,

$$\log_{10} 0.03741 = 8 - 10 + 0.5730 = 8.5730 - 10.$$

14. The characteristic is 0.

$$7\left\{\begin{array}{l} 5\left\{\begin{array}{l} 0.8082 \\ 0.8087 \end{array}\right. \\ \quad 0.8089 \end{array}\right.$$

x	$\text{antilog}_{10}\ x$
0.8082	6.430
0.8087	?
0.8089	6.440

y (between 6.430 and ?), 0.010 (between 6.430 and 6.440)

$$\frac{y}{0.010} = \frac{5}{7}$$

$$y = \frac{5}{7}(0.010) \approx 0.007$$

$$\text{antilog}_{10}\ 0.8087 = 6.430 + 0.007 = 6.437$$

16. The characteristic is 2.

$$24\left\{\begin{array}{l} 19\left\{\begin{array}{l} 0.2601 \\ 0.2620 \end{array}\right. \\ \quad 0.2625 \end{array}\right.$$

x	$\text{antilog}_{10}\ x$
0.2601	1.820
0.2620	?
0.2625	1.830

y (between 1.820 and ?), 0.010 (between 1.820 and 1.830)

$$\frac{y}{0.010} = \frac{19}{24}$$

$$y = \frac{19}{24}(0.010) \approx 0.008$$

$$\text{antilog}_{10}\ 0.2620 = 1.820 + 0.008 = 1.828$$

Since the characteristic of the given logarithm is 2,

$$\text{antilog}_{10}\ 2.2620 = 1.828 \times 10^2 = 182.8.$$

18. The characteristic is 3.

$$37\left\{\begin{array}{l} 4\left\{\begin{array}{l} 0.0755 \\ 0.0759 \end{array}\right. \\ \quad 0.0792 \end{array}\right.$$

x	$\text{antilog}_{10}\ x$
0.0755	1.190
0.0759	?
0.0792	1.200

y (between 1.190 and ?), 0.010 (between 1.190 and 1.200)

$$\frac{y}{0.010} = \frac{4}{37}$$

$$y = \frac{4}{37}(0.010) \approx 0.001$$

$$\text{antilog}_{10}\ 0.0759 = 1.190 + 0.001 = 1.191$$

Since the characteristic of the given logarithm is 3,

$$\text{antilog}_{10}\ 3.0759 = 1.191 \times 10^3 = 1191.$$

20. The characteristic is 3 - 5 = -2.

$$10\left\{\begin{array}{l} 5\left\{\begin{array}{l} 0.6107 \\ 0.6112 \end{array}\right. \\ \quad 0.6117 \end{array}\right.$$

x	$\text{antilog}_{10}\ x$
0.6107	4.080
0.6112	?
0.6117	4.090

y (between 4.080 and ?), 0.010 (between 4.080 and 4.090)

$$\frac{y}{0.010} = \frac{5}{10}$$

$$y = \frac{5}{10}(0.010) = 0.005$$

(continued)

$$\begin{aligned}&\text{antilog}_{10}\ 0.6112\\ &= 4.080 + 0.005\\ &= 4.085\end{aligned}$$

Since the characteristic of the given logarithm is -2,

$$\text{antilog}_{10}(3.6112 - 5) = 4.085 \times 10^{-2} = 0.04085.$$

22. The characteristic is 20 - 22 = -2.

$$5\left\{\begin{array}{l} 1\left\{\begin{array}{l}0.9978\\0.9979\end{array}\right.\\ \ \ 0.9983\end{array}\right. \quad \begin{array}{c} x \\ \hline \end{array} \quad \begin{array}{c}9.950\\ ?\\ 9.960\end{array} \left.\begin{array}{l} y\}\\ \ \end{array}\right\}0.010$$

x	$\text{antilog}_{10}\ x$
0.9978	9.950
0.9979	?
0.9983	9.960

(Differences: 1 and 5 in x; y and 0.010 in $\text{antilog}_{10}\ x$.)

$$\frac{y}{0.010} = \frac{1}{5}$$

$$y = \frac{1}{5}(0.010) = 0.002$$

$$\begin{aligned}&\text{antilog}_{10}\ 0.9979\\ &= 9.950 + 0.002\\ &= 9.952\end{aligned}$$

Since the characteristic of the given logarithm is -2,

$$\text{antilog}_{10}\ 20.9979 - 22 = 9.952 \times 10^{-2} = 0.09952.$$

24. The characteristic is 7 - 10 = -3.

x	$\text{antilog}_{10}\ x$
0.7396	5.490
0.7397	?
0.7404	5.500

(Differences: 1 and 8 in x; y and 0.010 in $\text{antilog}_{10}\ x$.)

$$\frac{y}{0.010} = \frac{1}{8}$$

$$y = \frac{1}{8}(0.010) \approx 0.001$$

$$\begin{aligned}&\text{antilog}_{10}\ 0.7397\\ &= 5.490 + 0.001\\ &= 5.491\end{aligned}$$

Since the characteristic of the given logarithm is -3.

$$\text{antilog}_{10}\ 7.7397 - 10 = 5.491 \times 10^{-3} = 0.005491.$$